HYDROGEL POLYMERS: SYNTHESIS AND CHARACTERIZATION

HYDROGEL POLYMERS: SYNTHESIS AND CHARACTERIZATION

By

Prof. Alias Smith

2016

SBS Publishers & Distributors Pvt. Ltd.
New Delhi

ISBN 13 : 9789380090900

First Published in 2016

Published by:

SBS PUBLISHERS & DISTRIBUTORS PVT. LTD.

2/9, Ground Floor, Ansari Road, Darya Ganj,

New Delhi - 110002,

INDIA

Tel: 0091.11.23289119 / 41563911

Email: mail@sbspublishers.com

www.sbspublishers.com

Preface

Hydrogel products constitute a group of polymeric materials, the hydrophilic structure of which renders them capable of holding large amounts of water in their three-dimensional networks. Extensive employment of these products in a number of industrial and environmental areas of application is considered to be of prime importance. As expected, natural hydrogels were gradually replaced by synthetic types due to their higher water absorption capacity, long service life, and wide varieties of raw chemical resources. The terms gels and hydrogels are used interchangeably by food and biomaterials scientists to describe polymeric cross-linked network structures. Gels are defined as a substantially dilute cross-linked system, and are categorized principally as weak or strong depending on their flow behavior in steady-state. Edible gels are used widely in the food industry and mainly refer to gelling polysaccharides. The term hydrogel describes three-dimensional network structures obtained from a class of synthetic and/or natural polymers which can absorb and retain significant amount of water. The hydrogel structure is created by the hydrophilic groups or domains present in a polymeric network upon the hydration in an aqueous environment. This book reviews the preparation methods of hydrogels from hydrophilic polymers of synthetic and natural origin with emphasis on water soluble natural biopolymers. Recent advances in radiation cross-linking methods for the preparation of hydrogel are particularly addressed. Additionally, methods to characterize these hydrogels and their proposed applications are also reviewed.

Editor

Contents

Chapter 1

EFFECTS OF INORGANIC AND ORGANIC FERTILIZERS ON GROWTH AND PRODUCTION OF BROCOLI (BRASSICA OLERACEA L.)

Hala Kandil, Nadia Gad

Plant Nutrition Department, National Research Centre, Dokki, Egypt

ABSTRACT

A field experiment was conducted in research and production station, El- Nubaria location, National Research Centre, Egypt during winter season, 2008, to study the effect of different solution fertilizers formula and organic manure on vegetative growth, heads yield quantity and quality as well as nutrient composition of broccoli (Brassica oleracea var. italica). The obtained results showed that all mineral solution fertilizers gave a significant synergistic effect for broccoli growth, yield quantity and quality as well as nutrients composition compared the control (mineral N, P, K recommended fertilizers). The mineral formula 19: 19: 19 recorded the highest growth heads, yield and quality along with mineral content in broccoli. Using farmyard manure plus inorganic fertilizers enhanced all growth and yield parameters.

Applying farmyard manure plus the mineral solution fertilizer formula 19: 19: 19 caused the superior and optimum figures of broccoli growth, mineral composition as well as heads yield quantity and quality. Organic manure alone recorded the lowest one.

INTRODUCTION

Cruciferous vegetables are large and increasingly important vegetables. Broccoli, a member of the crucifer family of vegetables, is a rich source of sulforaphane, which has been shown to display potent ant carcinogenic properties. However over half of the world population fails to benefit from this because they lack a specific gene (GSTMI) that helps retain the compound in the body (Kirsh et al., 2007). Eating a few portions of broccoli each week may help to reduce the risk of cancer. The cancer- fighting properties of broccoli are not new and previous studies have related these benefits to the high levels of active plant chemicals called glucosinolates (Zhao et al., 2007). Eating more than one serving of broccoli a week reduces the risk of prostate cancer by up to 45 percent. Eating larger portions may also have additional benefits, since broccoli is also a rich source of many vitamins and minerals such as vitamin A and C, carotenoids, fibber, calcium and folic acid (Michaud et al., 2002).

Growing broccoli in the newly reclaimed soils is faced by various problems, such as cultivars, fertilization, low amounts of available nutrients and low organic matter content as well as poor hydrophilic, chemical and biological properties. Nkoa et al, (2002) found that using mineral fertilizer (N, P, K) increasing broccoli vegetative growth, yield and quality.

Organic manure play direct role in plant growth as a source of all necessary macro and micronutrients in available forms during mineralization and improving physical and chemical properties of soils (Chaterjee et al., 2005). Anant-Bahadur et al., (2006) pointed that organic matter plays an important role in the chemical behaviour of several metals in soils throughout its active groups (Flavonic and humic acids) which have the ability to retain the metals in complex and chelate forms.

The objective of this work is to evaluate vegetative growth, heads yield, heads quality and chemical content of broccoli plants as affected by different mineral fertilizers and organic matter.

MATERIALS AND METHODS

A field experiment was carried out to study the effect of different sources of mineral fertilizers and organic matter on broccoli. Soil analysis: Some physical and chemical properties of El-Nubaria soil, Research and Production station, National Research Centre are shown in Table 1.

Table 1. Some Physical and Chemical Properties of the Used Soil at El-Nubaria, Research and Production Station, National Research Centre

<table>
<tr><td>Soil property</td><td colspan="4">Particle size distribution %</td><td colspan="4">Soil moisture constant %</td></tr>
<tr><td rowspan="3">Physical</td><td>Sand</td><td>Silt</td><td>Clay</td><td>Texture</td><td>Saturation</td><td>FC</td><td>WP</td><td>AW</td></tr>
<tr><td rowspan="2">68.7</td><td rowspan="2">24.5</td><td rowspan="2">6.8</td><td rowspan="2">S L</td><td colspan="4">%</td></tr>
<tr><td>32.0</td><td>19.2</td><td>6.1</td><td>13.1</td></tr>
<tr><td rowspan="12">Chemical</td><td colspan="2">pH[a]</td><td colspan="2">EC[b] dS/m</td><td colspan="2">$CaCO_3$%</td><td colspan="2">OM [c]%</td></tr>
<tr><td colspan="2">7.8</td><td colspan="2">0.18</td><td colspan="2">7.07</td><td colspan="2">0.16</td></tr>
<tr><td colspan="4">Soluble cations (meq/l)</td><td colspan="4">Soluble anions (meq/l)</td></tr>
<tr><td>Ca^{++}</td><td>Mg^{++}</td><td>K^{+}</td><td>Na^{+}</td><td>$CO_3^{=}$</td><td>HCO_3^{-}</td><td>Cl^{-}</td><td>SO4</td></tr>
<tr><td>3.00</td><td>2.00</td><td>0.32</td><td>2.09</td><td>0.00</td><td>1.41</td><td>0.70</td><td>5.30</td></tr>
<tr><td colspan="2">Total</td><td colspan="2">Available</td><td colspan="4">Available micronutrients</td></tr>
<tr><td colspan="2">N</td><td>P</td><td>K</td><td>Fe</td><td>Mn</td><td>ZN</td><td>Cu</td></tr>
<tr><td colspan="4">mg/100 g soil</td><td colspan="4">Ppm</td></tr>
<tr><td colspan="2" rowspan="4">15.0</td><td rowspan="4">9.4</td><td rowspan="4">16.0</td><td>7.8</td><td>3.3</td><td>1.86</td><td>4.0</td></tr>
<tr><td colspan="4">Cobalt (ppm)</td></tr>
<tr><td>Soluble</td><td>Available</td><td colspan="2">Total</td></tr>
<tr><td>0.49</td><td>4.43</td><td colspan="2">15.00</td></tr>
</table>

a: Soil pH was measured in 1:2.5 soil-water suspension, b: EC was measured as dSm-1 in soil paste, S L: sandy loam

Particle size distributions along with soil moisture were determined as described by Blackmore (1972). Soil organic matter, CaCO3, EC, pH, cations and anions, soluble and available micronutrients were determined according to Black et al., (1982).

Experimental Work

A field experiment was conducted during successive season, 2008 at El-Nubaria farm, National Research Centre, El-Beheara governorate, Egypt to evaluate vegetative growth, heads yield, heads quality and chemical constituents of broccoli plants as affected by different mineral fertilizers and organic matter. Seeds of broccoli (Brassica oleraceae L.var. italica), Family crucifer were sown in the nursery in foam trays filled with a mixture of peat moss and sand (1:1 volume) on 1St of September in 2008 season. Seedlings were transplanted in the open field at 45 days age. The experiment contains 9 plots. Each plot area was 5 X 3 meter, consisting of three rows. Ten plants in each row (50 cm a part) were planted under drip irrigation system.

Fertilizers Treatments

- Recommended fertilizers (ammonium nitrate (33.5% N) at a rate 100 N unit/fed.; super phosphate calcium (15.5% P2O5) at a rate 60 unit P2O5/fed. and potassium sulphate (48% K2O) at a rate of 50 unit K2O/fed.).
- Solution fertilizer formula (10: 19: 40).
- Solution fertilizer formula (19: 19: 10).
- Solution fertilizer formula (19: 19: 19).
- Farmyard manure.
- Recommended fertilizers + farmyard manure.
- Solution fertilizer formula (10: 19: 40) + farmyard manure.
- Solution fertilizer formula (19: 19: 10) + farmyard manure.
- Solution fertilizer formula (19: 19: 19) + farmyard manure.

Table 2. Some chemical properties of farmyard manure

O.M %	Total N %	C/N ratio	pH (1:2.5)	EC dSm^{-1}	Available nutrients %		DTPA- extractable (ppm)			
					P	K	Fe	Mn	Zn	Cu
36.00	2.8	7.47	7.5	3.3	1.20	1.6	840	30.0	15.0	3.0

The quantities of the chemical fertilizers were split into three equal doses (3, 6 and 9 weeks after transplanting) beside plant. All the plants received natural agricultural practices whenever they needed.

Measurement of Plant Growth

After 60 days from transplanting, growth parameters were recorded i.e. plant high, pranches and leaves number per plant, leaf area as well as fresh and dry weight of both shoots and roots (3 plants from each treatment) were determined according to Gabal et al., (1984).

Measurement of Heads Yield

At mature stage, after 90 days from transplanting, all broccoli heads were harvested to record:- Head weight (g/plant), head high, head diameter, total heads yield (ton/fed) according to Gabal et al., (1984).

Measurement of Heads Quality

Chemical contents i.e. TSS, protein, total carbohydrates, vitamin (C), vitamin (A) were determined according to the method described by A.O.A.C. (1980) as well as nutrients content were determined according to Cottenie et al. (1982).

Statistical analysis of the obtained data was subjected to standard analysis of variance procedure. The values of LSD were calculated at 5% level according to Snedicor and Cochran (1982).

RESULTS AND DISCUSSION

Vegetative Growth Characters

Results in Table (3) showed that broccoli growth characters were significantly influenced by different mineral fertilizers. The highest plant height branches and leaves number/plant and leaves area were recorded by plants which supplied with formula (19: 19: 19) fertilizer. All solution mineral fertilizers can be arranged in decreasing order as follows: formula (19: 19: 19) > formula (19: 19: 10) > formula (10:19: 40) > N, P, K as recommended. The lowest values of broccoli growth parameterswere obtained by N, P and K as recommended (control).

Table 3 Vegetative growth of broccoli plants as affected by inorganic and organic fertilizer

Fertilizers addition	Plant high (cm)	Branches No/plant	Leaves No/ plant	Leaves area (cm^2)	Fresh weight (gm)		Dry weight (gm)	
					Shoot	Root	Shoot	Root
Without organic fertilizers								
Control-N, P, K Recommended	48.20	6	12	169	271	23.8	24.9	5.95
10: 19: 40	52.66	7	13	197	280	25.0	26.7	6.24
19: 19: 10	58.50	8	16	229	289	26.2	27.5	6.66
19: 19: 19	63.00	10	19	253	297	27.3	28.9	6.85
With organic fertilizers								
Farmyard manure	45.60	6	12	181	258	22.5	20.8	5.57
Control-N, P, K Recommended	53.81	7	14	212	291	27.7	25.2	6.92
10: 19: 40	57.52	8	17	247	302	30.1	26.5	7.53
19: 19: 10	64.66	10	20	279	313	31.7	27.4	7.89
19: 19: 19	67.55	12	25	294	322	32.6	28.7	8.18
LSD 5%	0.30		-	2.80	1.62	0.31	0.28	0.07

Data also indicated that, it is worthy to notice the positive effect of organic manure fertilizer. It is obvious that farmyard manure had a promotive effect on all growth parameters include fresh and dry weights of broccoli shoots and roots. Data in Table 3 also, indicated that great differences in plant height were occurred under applied treatments. The values of plant height ranged from 48.20 cm for control to 63.00 cm for formula (19: 19: 19) treatment. On the other hand, number of branches and leaves/plant slightly different among treatments. The same trend took place for both fresh and dry weights of shoots and roots. Control plants had the lowest value against the highest one for formula (19: 19: 19) fertilizer. The values of fresh shoots and roots were 271 and 23.8 g/plant under control treatment versus 297 and 27.3 g/plant under treatment of formula (19: 19: 19) fertilizer. Regarding dry weights shoots and roots, data took a similar trend. The values reached

24.9 and 5.95 g/plant for control and 28.9 and 6.85 g/plant for formula (19: 19: 19) fertilizer. It was noticed that mineral solution fertilizers which used alone had values lower than that combined farmyard manure (FYM). Data in Table (3) clearly indicated that the vital role of the importance (FYM) as a source for organic nitrogen. These results are in agreement with those obtained by Rakesh et al., (2006) who showed that organic manure plus mineral fertilizer increase vegetative growth of broccoli plants.

Head Yield of Broccoli

The presented data in Table (4) concerning with the yield parameters of broccoli as affected by different solution fertilizers of mineral and organic manure either alone or in combination.

Table 4. The heads yield of broccoli as affected by inorganic and organic fertilizer

Fertilizers addition	Head high (gm)	Head diameter (cm)	Head Weight (g/plant)	Head Weight (ton/fed.)			
				First	Second	**Third**	**Total**
Without organic fertilizers							
(control) N, P, K Recommended	13.38	14.56	191.22	1.67	1.72	1.53	4.92
10: 19: 40	13.84	15.80	198.60	1.92	1.80	1.59	5.31
19: 19: 10	14.56	16.04	211.00	2.06	1.98	1.69	5.73
19: 19: 19	15.20	16.21	229.72	2.93	2.07	1.84	6.84
With organic fertilizers							
Farmyard manure	11.68	13.51	133.50	1.49	1.27	1.07	3.83
(control) N, P, K Recommended	14.91	15.06	198.80	2.42	1.95	1.59	5.96
10: 19: 40	15.26	16.14	223.50	2.66	1.92	1.79	6.77
19: 19: 10	15.93	17.61	257.80	2.95	2.34	2.06	7.35
19: 19: 19	16.14	18.32	278.60	3.56	2.66	2.22	8.44
LSD 5%	0.16	0.18	0.27	0.06	0.02	0.04	0.09

All mineral solution fertilizers gave a significant synergistic effect for the yield parameters i.e. head high; head diameter, heads weight (g/plant) and heads weight (ton/fed.) of broccoli compared the control. The best treatment of broccoli yield components i.e. first, second and third harvests were recorded by using mineral solution fertilizers, formula (19: 19: 19) followed by formula (19: 19: 10) followed by formula (10: 19: 40).

Data in Table 4 showed using FYM alone gave the lowest broccoli heads yield. Addition of organic manure plus mineral fertilizer increase the heads total yield and its components of broccoli from 4.92 to 5.96 ton/fed. with NPK recommended, from 5.31 to 6.37 ton/fed. With formula (10: 19: 40), from 5.73 to 7.35 ton/fed. with formula

(19: 19: 10), finally, from 6.84 to 8.44 ton/fed. with formula (19: 19: 19). The results reveal, as expected and as mentioned by Chaterjee et al., (2005).

Chemical Constitute

Data presented in Table 5 pointed that the highest TSS, protein, Phenols, vitamin "A" and vitamin "C" in heads of broccoli was recorded by formula (19: 19: 19) mineral solution fertilizer. The effect of different mineral fertilizers on broccoli heads chemical contents (%) could be arranged in the following order: formula (19: 19: 19) > formula (19: 19: 10) > formula (10: 19: 40) > NPK recommended (control).

Application of FYM combined with mineral solution fertilizer formula (19: 19: 19) were the superior and optimum interaction treatments. It is worthy to notice that the combination between mineral solution fertilizer formula (19: 19: 19) and FYM, caused the highest increase in all chemical constituents (TSS, protein, Phenols, vitamin "A" and vitamin "C").

Table 5 Chemical constituents of broccoli heads as affected by inorganic and organic fertilizer

Fertilizers addition	TSS	Protein	Phenols	Vitamin "A"	Vitamin "C"
	(%)			mg/ 100 gm fresh tissue	
Without organic fertilizers					
(control) N, P, K Recommended	7.18	21.40	6.86	15.35	77.80
10: 19: 40	7.67	22.96	7.54	15.90	81.92
19: 19: 10	7.96	23.84	7.98	16.72	85.61
19: 19: 19	8.21	25.52	8.35	17.06	92.35
With organic fertilizers					
Farmyard manure	6.13	19.47	6.24	14.86	72.60
(control) N, P, K Recommended	7.56	24.77	7.94	16.11	93.89
10: 19: 40	8.97	25.08	8.30	17.19	99.78
19: 19: 10	8.53	26.77	8.87	17.78	104.09
19: 19: 19	8.89	27.77	9.33	18.00	108.20
LSD 5%	0.09	0.29	0.06	0.17	0.38

The combination between FYM and mineral solution fertilizer formula (19:19:10) occupied the second order. While the combination between FYM and mineral solution fertilizer formula (10:19:40)

occupied the third order. Finally, the recommended N, P, K fertilizers with FYM occupied the fourth order. The lowest values of chemical constituents were recorded by organic manure alone. These data are in harmony with those obtained by Nanwai et al., (1998) how found that, the supplying vegetable crops with organic and inorganic fertilizers was proved to be very essential for producing higher yield and its quality.

Mineral Composition

Table 6. Minerals composition of broccoli heads as affected by inorganic and organic fertilizer

Fertilizers addition	Macronutrients (%)			Micronutrients (ppm)			
	N	P	K	Fe	Mn	Zn	Cu
Without organic fertilizers							
(control) N, P, K Recommended	3.43	0.28	1.89	129	48.92	33.55	28.50
10: 19: 40	3.68	0.31	1.97	132	51.03	35.20	31.02
19: 19: 10	3.82	0.36	2.14	137	53.61	37.14	34.50
19: 19: 19	4.09	0.42	2.23	141	55.52	39.66	36.72
With organic fertilizers							
Farmyard manure	3.12	0.26	1.80	126	44.60	30.41	26.0
(control) N, P, K Recommended	3.97	0.39	2.39	139	48.91	39.51	32.81
10: 19: 40	4.02	0.45	2.45	144	51.78	42.37	34.60
19: 19: 10	4.29	0.49	2.51	151	54.21	45.13	37.00
19: 19: 19	4.45	0.54	2.59	156	57.34	47.50	39.44
LSD 5%	0.134	0.085	0.033	1.715	0.990	0.281	0.462

Data in Table 6 demonstrate the influence of different mineral fertilizers and organic manure on minerals content in heads of broccoli. It is clear that mineral solution fertilizer formula (19: 19: 19) exhibited the highest values of all nutrients under study i.e. N, P, K, Mn, Zn, Cu. Control treatment which received the recommended doses of mineral fertilizers of N, P and K only showed the minimum content values for all elements. Therefore, using organic manure with mineral fertilizers achieved the best figures either on macronutrients or micronutrients. These results agree with those obtained by El-Shakry (2005) who found that organic manure enhances nutrients absorption by sweet fennel plants.

CONCLUSION

Mineral nitrogen from mineral solution fertilizers represent the easier available from of nitrogen compare recommended mineral fertilizers. Using organic manure plus inorganic solution fertilizers gave a significant promotive effect of plant growth, heads yield, chemical constituents and mineral composition of broccoli. Organic manure enhances soil aggregation, aeration, water holding capacity and amended the root system by slow release flow of nutrients which in combination creates favourable conditions for root respiration, nutrients absorption, root and upper parts growth and yield quantity and quality. Organic manure increasing the fertility and productivity of sandy soils.

REFERENCES

1. A.O.A.C. 1980, Official Methods of Analysis of The Association of Official Edition, Washington, D. C.
2. ANANT - BAHADUR, JAGDISH- SINGH, K.P. SINGH, A.K. UPADHYAY,
3. MATHURA-RAI, 2006, Effect of organic amendments and biofertilizers on
4. growth, yield and quality attributes of Chinese cabbage (Brassica pekinensis). Indian J. of Agric. Sci., 76 (10):596- 598.
5. BLACK, C.A.; D.D. EVANS; L.E. ENSMINGER; G.L. WHITE AND F.E. CLARCK, 1982, Methods of soil analysis Part 1 and 2, Agron. Inc. Madison. Wisc.
6. BLACKMORE, L.C., 1972, Methods for chemical analysis of soils, New Zealand soil Durean, P.A2 1, Rep. No. 10.
7. CHATERJEE, B.; P. GHANTI; U. THAPA AND P. TRIPATHY, 2005, Effect of organic nutrition in spro broccoli (Brassica aleraceae var. italica plenck),
8. Vegetable Science, 33 (1): 51-54.
9. COTTENIE, A.; M. VERLOO; L. KIEKENS; G. VELGH AND R. CAMERLYNCK, 1982, Chemical Analysis of Plant and Soil. Lab, Anal. Agrochem. State Univ. Gent, Belgium, 63.
10. EL-SHAKRY, M.F.Z., 2005, Effect of bio fertilizers, nitrogen sourced and levels on vegetative growth characters, yield and quality and yield oil of sweet fennel. Ph.D.Thesis, Fac. Agric. Cairo Univ. Egypt.
11. GABAL, M.R.; I.M. ABD-ALLAH; F.M. HASS AND S. HASSANNEN, 1984,
12. Evaluation of some American tomato cultivars grown for early summer production in Egypt, Annals of Agric. Sci. Moshtohor, 22:487-500.
13. KIRSH, V.A.; U. PETERS; S.T. MAYNE; A.F. SUBAR; N. CHATTERJEE; C.C.

14. JOHNSON AND HAYES, 2007, Prospective study of fruit and vegetable intake and risk of prostate cancer, Journal of the National Cancer Institute. Published online a head of print doi: 10. 1093/ jnci/ djm 065.

15. MICHAUD, D.S.; P. PIETINEN; P.R. TAYLOR; M. VIRTANEN; J. VIRTAMO AND D. ALBANES, 2002. Intates of fruits and vegetables, carotenoids and vitamin A, E, C in relation to the risk of bladder cancer in the ATBC cohort study, Br. J. Cancer, 87: 960- 965.

16. NANWAI, R.K.; B.D. SHARMA AND K.D. TANEJA, 1998, Role of organic and inorganic fertilizers for maximizing wheat (Triticum) aestivum yield in sandy loam soils., Crop research, Hisar. 16 (2): 159- 161.

17. NKOA, R.; J. COULOMBE; Y. DESJARDINS; J. OWEN AND N. TREMBLAY, 2002, Nitrogen supply phasing increases broccoli (Brassica aleraceae var. italica) growth and yield, Acta-Horticulture (571): 163-170.

18. RAKESH, S.; S.N.S. CHAURASIA AND S.N. SINGH, 2006, Response of nitrogen sources and spacing on growth and yield of broccoli (Brassica aleraceae var.

19. italica plenck), Vegetable Sci. 33 (2): 198- 200.

20. SNEDCOR, G. W. AND W. G. COCHRAN, 1982, Statistical methods, 7th Edition Iowa State Univ. Press. Ames. Iowa, USA.

21. ZHAO, H.; J. LIN; H. BARTON GROSSMAN; L.M. HERNANDEZ; C.P. DINNEY AND X. WU, 2007, Dietary isothiocyanates, GSTMI, GSTTI, NAT2 polymorphisms and bladder cancer risk, International Journal of Cancer, 120 (10): 2208- 2213.

Chapter 2

INFLUENCE OF ALTERNATIVE ORGANIC FERTILIZERS ON THE ANTIOXIDANT CAPACITY IN HEAD CABBAGE AND CUCUMBER

Bimova PAVLA[1], Robert POKLUDA[1]

[1]Mendel University of Agricultural and Forestry in Brno, Faculty of Horticulture, Department of Vegetable Production and Floriculture, Valtická 337, 691 44 Lednice; Czech Republic

ABSTRACT

Conventional mineral fertilizer was compared with alternative organic fertilizers for the crop of head cabbage and cucumber. There were seven different treatments: Agormin (an organo-mineral fertilizer), Agro (made from poultry bedding and molasses), conventional farmyard manure, compost, Dvorecky agroferm (dried, aerobically-fermented farmyard manure), mineral fertilizer, and an unfertilized control. All treatments were applied at rates providing approximately the same level of nutrients. The level of the total antioxidant capacity (TAC) was measured by FRAP assay immediately after harvest. Average value of TAC in fresh cabbage was 23660 mg GA.100 g^{-1} in the year 2005 and 29527 mg GA.100 g^{-1} in the year 2006. Average value of TAC in field cucumber was 12532 mg

GA.100 g^{-1} in the year 2005 and 10460 mg GA.100 g^{-1} in the year 2006. This study shows that alternative, organic fertilizers have similar or even better positive effects than farmyard manure and that they can contribute to the improvement of the nutritional value of vegetable production.

INTRODUCTION

In the past, agricultural production was focused on maximizing the quantity of fruits and vegetables produced for commercial market. However, modern consumers are now interested in optimizing the nutritional composition of foods. Therefore, much attention has now been placed on the agricultural practices that will enhance the nutritional content of fruit and vegetable being produced today (Wang, 2006). Fruit, vegetables, nuts and seeds, provides a rich source of antioxidant vitamins, and other phytochemicals with antioxidant characteristics. An antioxidant may be defined as any substance that when present at low concentrations, compared with those of the oxidizable substrate, significantly delays or inhibits oxidation of that substrate (Antolovich et al., 2002). The antioxidant content of fruits and vegetables may contribute to the protection they offer from disease. Because plant foods contain many different classes and types of antioxidants, knowledge of their total antioxidant capacity (TAC), which is the cumulative capacity of food components to scavenge free radicals, would be useful for epidemiologic purposes (Pellegrini, 2003; Lindsay and Astley, 2002). Antioxidant capacity has been assessed in many ways and there is large variety of ways and results. The total antioxidant capacity assays by 3 different methods were studied by Pellegrini, (2003). Table 1 shows different method with different results, which used different units.

Cruciferous vegetables, including cabbage (Brassica oleracea convar. capitata var. capitata), have a high nutritional value and contain organo-sulphur phytochemicals that increase their antioxidant capacity, which may have ant carcinogenic effects (Kim et al., 2004; Kurilich at al., 1999). The average value of TAC presented by Zloch, (2004) was 97 mg GA.100 g-1 in cabbage and 27 mg GA.100 g^{-1} in cucumber. Head cabbage and cucumber are the most important field vegetable crops in the Czech Republic, as well as in many other countries.

Cultivation of cabbage and cucumber demands for organic fertilization made by farmyard manures (Din et al., 2007; Murison and Napier, 2006). Farmyard manure is natural source of organic matter and vegetable production requires continuous applications of organic compounds (Bunting, 1965). However, a lot of farms specialize in vegetable production these days and they have no animals, so traditional farmyard manure is consequently in short supply. On the other hand, the farms specialized in rearing livestock have the opposite problem, namely, an abundance of manure which is difficult to dispose of. This surplus farmyard manure can be returned to the soil by processing it to make an organic fertilizer by aerobic fermentation and drying (Debosz et al. 2002).

Table: 1 Values of TAC in vegetables presented by Pellegrini et al., (2003)

Assays methods	FRAP[a]	TRAP[b]	TEAC[c]
	(mmol Fe^{2+}.kg^{-1} f. m.)	(mmol Trolox.kg^{-1} f. m.)	
Cucumber	0.71	-	0.43
Cabbage, white	5.79	2.83	1.15

[a]FRAP = Ferric reducing antioxidant power (mmol Fe^{2+}.kg^{-1} f. m.)
[b]TRAP = Total radical-trapping antioxidant parameter (mmol Trolox.kg^{-1} f. m.)
[c]TEAC = Trolox equivalent antioxidant (mmol Trolox.kg^{-1} f. m.)

Table 2 Characteristics of applied fertilizers

Fertilizer (company)	Compounds	N (%)	P (%)	K (%)	Dose (t.ha^{-1})
Agormin (AGRO CS a.s., CZ)	peat, basic macroelements	3.7	1.4	7.1	2.5
Agro (MeMon B.V., NL)	poultry bedding, molasses	3.5	1.5	7	1
Dvorecký Agroferm (Agropodnik Dvorce a.s., CZ)	fermented and dried cow-dung	1.6	0.6	1.0	0.8-1
Farmyard Manure (local source - Lednice, CZ)	treated mixture – bedding, stiff and liquid feaces of livestock	3.0	1.4	2.7	55 cabbage 40 cucumber

Mineral Fertilizer (AGRO CS a.s., CZ)	ammonium sulphate	21	-	-	
	potassium sulphate	-	-	50	
	superphosphate	-	18	-	
Horticultural compost (AGRO CS a.s., CZ)	plant waste with added dolomitic calcite	1.0	0.4	1.7	30

The aim of this study was to observe the effect of these alternative organic fertilizers on the total antioxidant capacity (TAC) of head cabbage and cucumber.

MATERIALS AND METHODS

The two-year experiment took place at the Faculty of Horticulture of Mendel University of Agriculture and Forestry Brno, in Lednice in 2005 and 2006. There were established 7 variants with following treatments: Agormin (an organo-mineral fertilizer), Agro (an organic fertilizer), control (unfertilized), conventional farmyard manure, conventional mineral fertilizer, Dvorecký agroferm (an organic fertilizer), horticultural compost. More detailed characteristic of applied fertilizers is displayed in Table 2. Each treatment was provided with three replicates.

The application rates were in accordance with the manufacturers' guidelines and in the case of farmyard manure as recommended by Malý et al., (1998) (Table 2). All variants, except the control one, were fertilized on the same level of nutrients according to the soil analysis and supposed yield (1 ton of cabbage uptake is as follows: 3.57 kg N; 0.57 kg P; 3.57 kg Kand 1 ton of cucumber uptake is as follows: 1.67 kg N; 0.70 kg P; 2.33 kg K). Corrections were made depending on the organic fertilizer, the preceding crop and the content of nutrients in the soil (Hlušek, 1996).

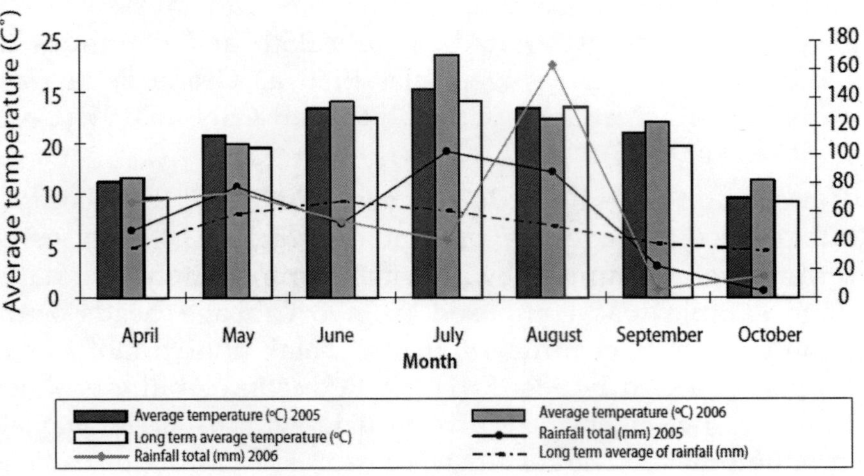

Figure 1. Climatic condition

Table 3 Mean value of TAC in cabbage in the years 2005 and 2006

Fertilizer	TAC in 2005(mg GA.100 g−1)		TAC in 2006 (mg GA.100 g−1)	
Agormin	292 ± 16	ab	292 ± 21	a
Agro	212 ± 43	ab	295 ± 19	a
Hort. Compost	201 ± 34	ab	284 ± 22	a
Control	233 ± 32	ab	299 ± 27	a
Dvorecký agroferm	314 ± 30	b	316 ± 52	a
Farmyard manure	211 ± 65	ab	296 ± 32	a
Mineral fertilizer	186 ± 73	a	281 ± 5	a
Mean	236 ± 60		295 ± 27	

The cultivar of head cabbage used was 'Trvalo F1' (Semo Smržice, CZ), which is acceptable for long-term storage. Harvesting was done on October 11th 2005 and on October 20th 2006 and the heads were classified as Grade I or II quality in accordance with local norm ČSN 463113 (UNECE STANDARD FFV-09). The cultivar of cucumber used was 'Linda F1 MIX' (Semo Smržice, CZ), which is acceptable for field cultivation. Sequential harvesting (7-days intervals) started on August 9th and finished on September 22nd in the year 2005. Following year harvest started on July 25th and continues until September 4th. The fruits were classified as Grade extra class, I or II quality in accordance with local norm ČSN 463155 (UNECE STANDARD FFV-15).

The climatic conditions during years 2005 and 2006 compared with long-term averages are presented in Figure 1. Analyses were done from average samples by measuring immediately after harvest. Average sample was composed from 3 cabbage heads without stalk and from 5 cucumbers fruits. Total antioxidant capacity (TAC) was assessed by FRAP (Ferric Reduction Ability of Plasma) assay according Zloch (2004) using a Jenway 6100 (Jenway, UK) spectrophotometer. The results were expressed as mg of gallic acid equivalents per 100 g of fresh matter (mg GA.100g^{-1}). All the results were processed by ANOVA and Tukey's test using the statistical program Unistat 5.1 (Unistat, USA) at probability 95%.

RESULTS AND DISCUSSION

Cabbage

The average value of TAC in cabbage was 236 ± 60 mg GA.100 g^{-1} in year 2005 and 295 ± 27 mg GA.100 g^{-1} in year 2006 immediately after harvest. The mean values of TAC in cabbage heads for each treatment are shown in Table 3. The results show threefold higher value of TAC in winter cabbage in comparison with data, which mentioned Zloch, (2004). The highest mean levels of TAC were observed in Dvorecký agroferm, and were significantly higher than those of the mineral fertilizer treatment in 2005. There were no significant differences between the treatments in levels of TAC in the year 2006, however the highest value of TAC was found in Dvorecký agroferm (Figure 2). According to Ismail et al., (2004) the variation between years is due to environmental factors, such as a climatic growth conditions. The different climatic conditions of years 2005 - 2006 could probably influence these results.

Cucumber

The average value of TAC in cucumber was 125 ± 32 mg GA.100 g^{-1} in year 2005 and 114 ± 57 mg GA.100 g^{-1} in year 2006. The mean value of TAC in cucumber for each treatment is shown in Table 4. The results show three as far as fivefold higher value of TAC in cucumber in comparison with data of Zloch (2004). The differences of the results obtained from this study compared to the literature may have been due to the differences in genotype of the cabbage and cucumber used or due to the different climate conditions. The highest mean levels of TAC were observed in Dvorecký agroferm (2005) and control (2006) however control and Agormin recorded the lowest value of TAC in 2005. The variation between 2 years is due to environmental factors. As show figure 1, years 2005 and 2006 were really different as far as average temperature and total rainfall. There were no significant differences between the treatments, in the year 2005. Agormin, Dvorecký agroferm and control shown significantly higher TAC in cucumber than compost, farmyard manure and mineral fertilizer, in the year 2006.

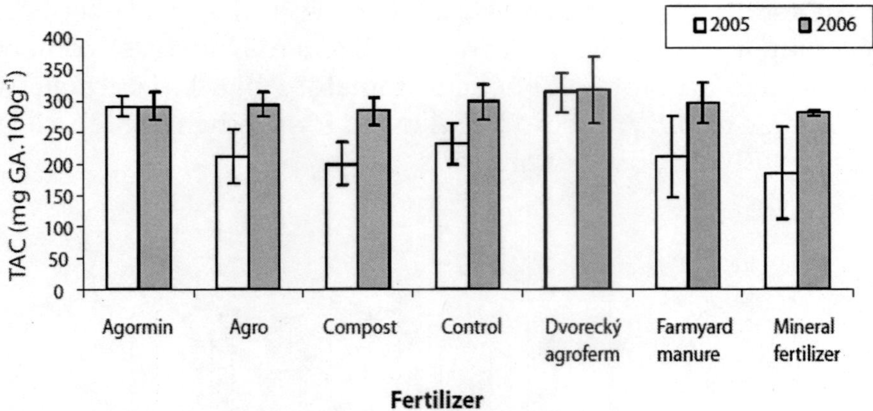

Figure 2 Effect of applied fertilizers on TAC (mg GA.100 g^{-1}) in head cabbage (2005-2006), 95% interval of significance level

Table 4 Average value of TAC in cucumber in the years 2005 and 2006

Fertilizer	TAC in 2005(mg GA.100 g^{-1})		TAC in 2006(mg GA.100 g^{-1})	
Agormin	103 ± 50	a	175 ± 11	b
Agro	141 ± 23	a	111 ± 4	ab
Hort. compost	133 ± 10	a	60 ± 38	a
Control	108 ± 14	a	184 ± 1	b
Dvorecký agroferm	142 ± 44	a	157 ± 14	b
Farmyard manure	129 ± 32	a	62 ± 12	a
Mineral fertilizer	117 ± 43	a	53 ± 33	a
Mean	125 ± 32		114 ± 57	

Means are followed by standard deviation; Different letters indicate significant differences at P≤ 0.05 (Tukey's HSD test)

Many experiments have shown that cabbage gives a good response to manure and most of earlier experiments indicate a superiority of manure over commercial fertilizers (Din et al., 2007). The studies by Abou El-Magd et al. (2006) and Toor et al., (2006) are quoted in support of nutritional benefit of organic fertilizers. According to Toor, (2006) organic fertilizers may increase content of ascorbic acid and total phenolic in tomato. Abou El-Magd showed that organic fertilizers could be followed for producing high yield of broccoli with high quality of heads.

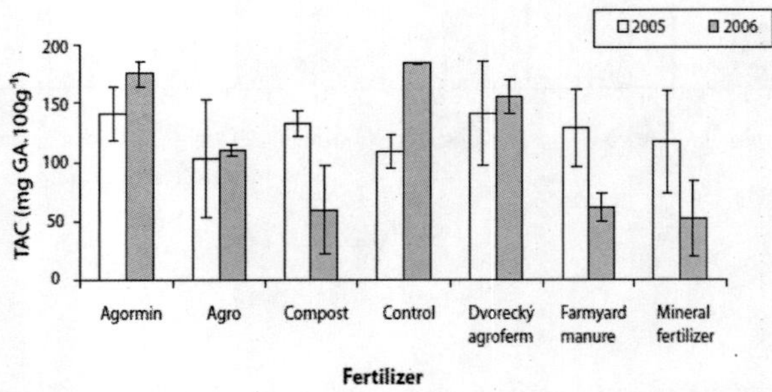

Figure 3 Effect of applied fertilizers on TAC (mg GA.100 g^{-1}) in cucumber (2005-2006), 95% interval of significance level

Table 5 Order of TAC in cabbage and cucumber during 2 years

Treatment	Agormin	Agro	Control	Dvorecký agroferm	Farmyard manure	Mineral fertilizer	Hort. compost
Cabbage	7	9	5	2	7	14	12
Cucumber	9	6	7	4	9	12	9

For simplified evaluation of cucumbers and cabbages results was used Table 5, which showed the order of treatments according to value of TAC. There were 7 different treatments, consequently treatment with highest value was classified as number 1 and the treatment with lowest value of TAC had number 7. Theoretical value of the best treatment is 2 and the worst is 14, because biennial records are presented. This table showed that the best treatment was Dvorecký agroferm and the worst one was mineral fertilizer. Farmyard manure was worse than control and other alternative organic fertilizers (Agro, Agormin) showed similar or the same results as well as farmyard manure.

CONCLUSIONS

The finding of this study indicate that different fertilizers contributed to affecting of TAC and so affected nutritional value of vegetable. Alternative organic fertilizers can positive influenced TAC in vegetable. Also their effect to nutritional value can be positive. Alternative, organic fertilizers as well as Dvorecký agroferm and Agormin have similar or even better qualities as farmyard manure, so they can contribute to the improvement of nutritional value of vegetable production.

ACKNOWLEDGEMENT

Work was supported by project of Ministry of Agriculture of Czech Republic No. QF4195.

REFERENCES

1. Abou El-Magd, M. M., A. M. El-Bassiony, Z. F. Fawzy, 2006, Effect of Organic Manure with or Without Chemical Fertilizers on Growth. Yield and Quality of Some Varieties of Broccoli Plants. Journal of Applied Sciences Research 2 (10), 791-798.
2. Antalovich M. et al., 2002, Methods for testing antioxidant activity. Analyst 127, 183–198.
3. Bunting, A. H., 1965, Effects of organic manures on soils and crops. Proceedings of the Nutrition Society 24 (1), 29–38.
4. Debosz, K. et al., 2002, Evaluating effects of sewage sludge and household compost on soil physical, chemical and microbiological properties. Applied Soil Ecology 19 (3), 237-248.
5. Din, M., M. Qasim, M. Alam, 2007, Effect of different levels of N, P and K on the growth and yield of cabbage. Journal of Agricultural Research 45 (2), 171-176.
6. Hlušek J., 1996, Základy výživy a hnojení zeleniny a ovocných kultur, Praha, Institut výchovy a vzdělávání Mze ČR: 48.
7. Ismail A., Z. M. Marjan, C. W. Foong, 2004, Total antioxidant activity and phenolic content in selected vegetable. Food Chemistry 87 (4), 581-586.
8. Kim, D. O., O. I. Padilla-Zakour, P. D. Griffiths, 2004, Flavonoids and Antioxidant Capacity of Various Cabbage Genotypes at Juvenile Stage. Journal of Food Science 69 (9), 685-689.
9. Kurilich, A. C. et al., 1999, Carotene, tocopherol, and ascorbate contents in subspecies of Brassica oleracea. Journal of Agricultural and Food Chemistry 47 (4).
10. Kurilich, A. C. et al., 2002, Antioxidant Capacity of Different Broccoli (*Brassica oleracea*) Genotypes Using the Oxygen Radical Absorbance Capacity (ORAC) Assay. Journal of Agricultural and Food Chemistry 50, 5053-5057.
11. Lindsay, D. G., S. B. Astley, 2002, European research on the functional effects of dietary antioxidants. Eurofeda, Molecular Aspects of Medicine, 23, 1-38.
12. Malý, I. et al., 1998, Polní zelinářství. Praha: Agrospoj.
13. Murison, J., T. Napier, 2006, Cabbage growing, Primefact 90, NSW DPI, State of New South Wales, 7.
14. Pellegrini, N. et al., 2003, Total assay antioxidant capacity of plant foods, beverages and oils consumed in Italy assessed by three different in vitro. Journal of nutrition 133 (9), 2812-2819.
15. Toor, R. K. et al., 2006, Influence of different types of fertilisers on the major antioxidant components of tomatoes, Journal of Food Composition and Analysis, volume 19, Issue 1, 20-27, ISSN: 0889-1575.
16. Wang, S. Y., 2006, Effect of Pre-harvest Conditions on Antioxidant Capacity in Fruits. Acta Horticulturae 712, 299-306.
17. Zloch, Z., J. Čelakovský, A. Aujezdská, 2004, Stanovení obsahu polyfenolů

a celkové antioxidační kapacity v potravinách rostlinného původu, on-line, Ústav hygieny Lékařské fakulty UK, Plzeň, Cited, 2004-12-10, http://www.danone-institut.cz/files/2004.04/.

Chapter 3

MICROBIOLOGICAL SAFETY OF CHICKEN LITTER OR CHICKEN LITTER-BASED ORGANIC FERTILIZERS: A REVIEW

Zhao Chen [1] and Xiuping Jiang [2,]

[1]Department of Biological Sciences, Clemson University, SC 29634, USA; E Mail: zchen5@clemson.edu

[2]Department of Food, Nutrition, and Packaging Sciences, Clemson University, SC 29634, USA

ABSTRACT

Chicken litter or chicken litter-based organic fertilizers are usually recycled into the soil to improve the structure and fertility of agricultural land. As an important source of nutrients for crop production, chicken litter may also contain a variety of human pathogens that can threaten humans who consume the contaminated food or water. Composting can inactivate pathogens while creating a soil amendment beneficial for application to arable agricultural land. Some foodborne pathogens may have the potential to survive for long periods of time in raw chicken litter or its composted products after land application, and a small population of pathogenic cells may even regrow to high levels when the conditions are favorable for growth. Thermal processing is a good choice for inactivating pathogens in chicken litter or chicken litter-based organic fertilizers prior to land application. However, some populations may become acclimatized to a hostile environment during build-up or composting and develop heat resistance through cross-protection during subsequent high temperature treatment. Therefore, this paper reviews currently available information on the microbiological safety of chicken litter or chicken litter-based organic fertilizers, and discusses about further research on developing novel and effective disinfection techniques, including physical, chemical, and biological treatments, as an alternative to current methods

INTRODUCTION

Chicken litter is a mixture of feces, wasted feeds, bedding materials, and feathers [1,2]. Over 14 million tons of chicken litter is produced every year in the US, most of which is usually recycled and spread on arable land as a low cost organic fertilizer [3,4]. Poultry manure contains significant amounts of nitrogen because of the presence of high levels of protein and amino acids. Owing to its high nutrient content, chicken litter has been considered to be one of the most valuable animal wastes as organic fertilizer [5]. Chicken litter is also the source of human pathogens, such as *Salmonella,* Campylobacter *jejune,* and *Listeria monocytogenes,* that can potentially contaminate fresh produce or the environment and are frequently associated with foodborne outbreaks [1,6]. Composting of poultry waste prior to the application to agricultural land as an organic fertilizer is usually recommended to control pathogens in the end products. Nonetheless, several studies have demonstrated that some pathogenic cells have the potential to persist in the finished compost and also compost-amended soil [7–9]. Another major concern for composting is the possibility of pathogen regrowth [10], indicating that a small population of pathogen that either survives the composting process or gets transferred from the environment may multiply to high populations under favorable conditions.

Currently, physical heat treatment (heat-drying after composting or without composting) is one of the most commonly applied techniques to reduce or eliminate potential pathogens in animal wastes [1,2]. The physically heat-treated chicken litter is recommended and used by produce growers. However, some pathogenic cells may have the potential to become acclimatized to the hostile environment during build-up or composting, cross-protecting them against subsequent high temperature treatment [11,12]. Therefore, some current guidelines for heat-treated animal wastes may not be sufficient to eliminate pathogens from the physically heat-treated chicken litter as organic fertilizer. Land spreading of contaminated chicken litter or chicken litter-based organic fertilizers (fertilizers derived from chicken litter sources) can also potentially lead to the introduction of foodborne pathogens into the food chain. Contamination of fresh produce with fecal pathogenic bacteria in the agricultural environment has been implicated as the main cause of numerous food poisoning outbreaks [13]. Therefore, to ensure the absence of pathogens in the fresh chicken litter, poultry compost, or the physically heat-treated chicken litter, additional approaches such as physical, chemical, and biological treatments, should be considered as another means for pathogen control. Moreover, nutrient retention, fuel cost, efficiency, capital cost, and environmental and regulatory policies will be the principle factors when it comes to making decisions on selected processing techniques [14].

Although the application of poultry litter for commercial farming has rarely been associated with foodborne outbreaks, enhanced consumer awareness of food safety issues has increased the scrutiny of agricultural practices. This review thus focuses on the microbiological safety of chicken litter or chicken litter-based organic fertilizers.

PATHOGENS AND ANTIBIOTIC-RESISTANT BACTERIA IN CHICKEN LITTER OR CHICKEN LITTER-BASED

Organic Fertilizers

Chicken litter contains a large and diverse population of microorganisms. Microbial concentrations in chicken litter can reach up to 10^{10} CFU/g, and Gram-positive bacteria, such as *Actinomycetes*, *Clostridia/Eubacteria*, and *Bacilli/Lactobacilli*, account for nearly 90% of the microbial diversity [15]. Pathogens in chicken litter represent the major group of bacteria of special interest to litter processors.

A variety of pathogens can be found in chicken litter or chicken litter-based organic fertilizers, such as

Actinobacillus, Bordetalla, Campylobacter, Clostridium, Corynebacterium, Escherichia coli, Globicatella, Listeria, Mycobacterium, Salmonella, Staphylococcus, and *Streptococcus* [15–20]. While different microbes display different metabolic activities within the litter environment, high levels of background microflora may interfere with the survival and growth of pathogens in chicken litter. Fully understanding the levels and prevalence of human pathogens in chicken litter or chicken litter based-organic fertilizers is essential for developing intervention strategies for controlling produce contamination on farms. As shown in Table 1, microbiological surveys have revealed the prevalence of some foodborne pathogens in chicken litter or chicken litter-based organic fertilizers, depending on pathogen species and serotype, chicken age, season, geographic area, farm handling practice, and so on [19,21–23]. For example, Li *et al.* [23] observed that fecal samples of 18-week-old layer birds had the highest prevalence of *Salmonella* (55.6%), followed by the 25- to 28-wk birds (41.7%), 75- to 78-wk birds (16.7%) and 66- to 74-wk birds (5.5%). Renwick *et al.* [22] surveyed randomly selected commercial broiler chicken flocks in Canada to determine flock and management factors associated with the prevalence of *Salmonella* contamination in the floor litter. They found that the prevalence of *Salmonella* in floor litter samples was significantly associated with the age of the flock and the region of Canada in which the flock was located.

Active surveillance data on foodborne diseases from the United States reveal that among pathogens associated with foodborne outbreaks, *Salmonella, E. coli* O157:H7, *Campylobacter,* and *L. monocytogenes* are responsible for the majority of outbreaks. *Salmonella* spp. is the most widely distributed pathogen in chicken litter with poultry and eggs remaining as the predominant reservoir. During 1998–2008, foodborne disease outbreaks caused by *Salmonella* were associated most commonly with poultry meat products (30%) and eggs (24%) [24]. Chicken eggs can be contaminated with *Salmonella* either horizontally or vertically. The contamination of egg shell can result from horizontal transmission, such as fecal contact [25]. And vertical transmission of *Salmonella* has been observed in infected ovaries, oviducts, or infected eggs [26]. Although only low numbers of *Salmonella* can contaminate eggs via the fecal route, these small populations cannot be ignored. Notably, *S*. Enteritidis, *S*. Typhimurium, or *S*. Heidelberg present in chicken feces may not only penetrate into the interior of eggs but also multiply during storage [27]. *Salmonella* is more frequently isolated from chicken litter or fecal samples as compared to other pathogens being investigated and its prevalence level can range widely from 0 to 100%. And the population of *Salmonella* in chicken litter can range from 4 to 1.1×10^5 MPN/g litter [6].

Table 1. Prevalence of foodborne pathogens in chicken litter or chicken litter based-organic fertilizers.

Pathogen	Year/Location	Sample source	Sample type	Sample size	Prevalence	References
Actinobacillus	N.A. [a]/Canada	Broiler, hen, and turkey	Litter samples	44	2%	[16]
Campylobacter	1995/US	19 broiler flocks	Fecal samples	948	86%–100%	[19]
	1996–1997/US	Poultry	Litter samples intended for dairy cattle feed from 13 dairy ranches	104	- [b]	[28]
	2001/US	9 broiler flocks	Fecal samples	450	80%–100%	[19]
	N.A./Australia	28 sheds of 28 broiler farms	Litter samples	60 sites/shed and three sets of 20 were combined	36%	[6]
Clostridium	N.A./Canada	Poultry	Litter samples	44	57%	[16]
	N.A./Nigeria	Layer	Litter samples	N.A.	+ [c]	[20]
E. coli	1994–1995/US	Poultry	Litter samples	86 (64 composted, 18 not composted, 4 samples not analyzed)	- for *E. coli* O157:H7	[29]
	1996–1997/US	Poultry	Litter samples intended for dairy cattle feed from 13 dairy ranches	104	- for *E. coli* O157, 8%–15% for non-O157 *E. coli*	[28]
	N.A./Nigeria	Layer	Litter samples	N.A.	+	[20]
	N.A./Australia	28 sheds of 28 broiler farms	Litter samples	60 sites/shed and three sets of 20 were combined	100%	[6]
	2004–2007/US	Poultry	Samples of compost heaps with chicken litter or chicken carcasses	N.A.	26% surface and 6.1% internal samples (1st composting phase); absent in all samples (2nd composting phase)	[30]

Pathogen	Year/Location	Sample source	Sample type	Sample size	Prevalence	References
Listeria	N.A./Australia	28 sheds of 28 broiler farms	Litter samples	60 sites/shed and three sets of 20 were combined	-	[6]
	2004–2007/US	Poultry	Samples of compost heaps with chicken litter or chicken carcasses	N.A.	-	[30]
Mycobacterium	N.A./Canada	Poultry	Litter samples	44	5%	[16]
	N.A./Nigeria	Layers	Litter samples	N.A.	+	[20]
Salmonella	N.A./Canada	Poultry	Litter samples	44	7%	[16]
	N.A./US	Poultry from 5 premises	Litter samples	198	73%–89%	[31]
	1977/Canada	3 broiler flocks	Litter samples (top 1.27 to 2.54 cm layer)	N.A.	0%–2%	[32]
	1978–1979/Canada	60 broiler houses	Litter samples	15 from each house	30%	[33]
	1980–1981/Canada	Broiler	Litter and feces samples	36 and 2 for litter and feces samples, respectively	19%–89% and 0%–100% for feces and litter, respectively	[21]
	1989–1990/Canada	Broiler	Litter samples	12	76%	[22]
	1994–1995/US	Poultry	Litter samples (64 composted, 18 not composted, and no determination for 4 samples)	86	-	[29]
	1996–1997/US	Poultry	Litter samples intended for dairy cattle feed from 13 dairy ranches	104	-	[28]

Table 1. *Cont.*

Pathogen	Year/Location	Sample source	Sample type	Sample size	Prevalence	References
	2002/Nigeria	5 poultry farms	Fecal samples	120	38%	[34]
	2006–2007/Hungary	Broiler	Fecal samples	60	35%–45%	[35]
	N.A./US	Hen	Fecal samples	78	17%–56%	[23]
	N.A./Nigeria	Layer	Litter samples	N.A.	+	[20]
	N.A./US	7 broiler farms	Fecal samples	420	6%–39%	[36]
Salmonella	N.A./Australia	28 sheds of 28 broiler farms	Litter samples	60 sites/shed and three sets of 20 were combined	71%	[6]
	2004–2007/US	Poultry	Samples of compost heaps with chicken litter or chicken carcasses	N.A.	26% surface and 6.1% internal samples (1st composting phase); absent in all samples (2nd composting phase)	[30]
	N.A./Canada	Poultry	Litter samples	44	100%	[16]
Staphylococcus	1994–1995/US	Poultry	Litter samples (64 composted, 18 not composted, and no determination for 4 samples)	86	-	[29]
	N.A./Nigeria	Layers	Litter samples	N.A.	+	[20]
Streptococcus	N.A./Canada	Poultry	Litter samples	44	100%	[16]

N.A., not applicable; [b]-, no pathogen or selected microorganism was isolated; +, pathogen or selected microorganism was isolated.

E. coli is present in chicken litter with the prevalence rate as high as 100%; however, *E. coli* O157:H7 was not detected in chicken litter samples [28,29] or poultry compost samples [29,30]. The population of *E. coli* in reused chicken litter can reach up to 9.7×10^4 CFU/g while the population for single use litter has been found to be 4.2×10^5 CFU/g [6]. *Campylobacter*, followed by *Salmonella*, is the leading cause of bacterial gastroenteritis due to food consumption [37] and is also likely to be present in poultry wastes. The prevalence of *Campylobacter* in chicken litter or fecal samples can range from 0 to 100% and its average population level was reported to be *ca.* 10^5 CFU/g in fecal samples collected from broiler chicken flocks [19]. *L. monocytogenes* is usually absent (negative) in chicken litter and poultry compost, and this pathogen therefore appears not to be a significant issue in chicken litter or chicken litter-based organic fertilizers [6,30].

There are also growing concerns about the presence of antibiotic-resistant pathogens in animal manures from both on-farm exposure and off-farm contamination. Widespread dispersal of chicken litter or chicken litter-based organic fertilizers harboring antibiotic-resistant foodborne pathogens can be a serious environmental hazard. Furthermore, horizontal transfer of mobile antibiotic resistance genes from one bacterium to another can possibly occur under some conditions [38]. Nandi *et al.* [39] reported that Gram-positive bacteria were found to be the major reservoir of Class 1 antibiotic resistance integrin's in poultry litter. As antibiotics are routinely used for disease prevention and growth promotion, a low level of antibiotics may select antibiotic-resistance bacteria in the gastrointestinal tract of the animal and also under *in vitro* conditions when antibiotic-laden manure is applied to the agricultural land [40]. Table 2 lists the presence of antibiotic-resistant bacteria in chicken litter or chicken litter based-organic fertilizers, highlighting the need for better waste management practices by poultry producers.

The prevalence of some antibiotic-resistant bacteria in chicken litter or chicken litter based-organic fertilizers can reach more than 60% for selected microorganisms, while it should be noted that some bacteria, such as *E. coli*, *Enterococcus*, and *Providencia*, are found to be multi-resistant to various antibiotics. It is also known that the increased use of antibiotics in the poultry industry can introduce a selective pressure which leads to the development of resistance or even multi-resistance characteristics in some of the bacterial populations. Moreover, as was observed by Khan *et al.* [41], erythromycin-resistant *Staphylococci*, *Enterococci*, and *Streptococci* were only isolated from litter samples collected from poultry houses that had used the antibiotics. Isolation of antibiotic-resistant foodborne pathogens from chicken litter or chicken litter based-organic fertilizers raises concerns about possible transmission of these bacteria to fresh produce after land application since these pathogens can potentially transfer to the arable land from contaminated chicken litter or chicken litter-based organic fertilizers, and can also further contaminate surface and ground water through runoff. This suggests the poultry industry should follow prudent management options and safety precautions by establishing more effective disinfection guidelines to reduce the population of antibiotic-resistant pathogens and monitoring the potential infection of subsequent flocks with resistant bacteria. In addition, a judicious and moderate use of antibiotics may also help prevent the emergence of antibiotic resistance in pathogenic bacteria. In the meantime, it is of great significance to identify and characterize various isolated antibiotic-resistant pathogens from chicken litter or chicken litter based-organic fertilizers. Therefore, further research is warranted to evaluate the pathogenicity of these antibiotic-resistant isolates, as well as their persistence in manure-amended soil.

Table 2. Antibiotic-resistant bacteria in chicken litter or chicken litter based-organic fertilizers.

Pathogen	Year/Location	Sample source	Sample type	Sample size	Comments [b]	Reference
Coliforms	N.A. [a]/US	4 turkey farms (8 houses), 10 adult broiler breeder chicken farms (43 houses), and 30 broiler chicken farms (110 houses)	Litter samples	N.A.	In turkey litter, the percentage of NAL-resistant coliforms ranged from 0.6% to 61.9%. Two farms had houses containing coliforms resistant to ENR and SAR. There was also multiple resistance to AMP, TIO, CAM, KAN on all 4 turkey farms. There were no NAL-resistant isolates from any of the 10 adult broiler breeder chicken farms. All of the 30 broiler chicken farms with NAL-resistant isolates were also resistant to SAR.	[42]
E. coli	2004–2007/US	Poultry	Chicken litter, carcasses, pine shavings, pine fines, and fresh wood chips	30 compost samples of chicken litter and carcasses, 42 compost samples of chicken litter and pine shavings, 18 compost samples of chicken litter with pine fines, and 24 compost samples of chicken litter, carcasses, and fresh wood chips	Isolates from California chicken litter/horse track had higher levels (63%) of resistance to AMP as compared with poultry compost in South Carolina (0%). *E. coli* isolates from poultry composts on South Carolina farms were found to be more resistant to TET (50%) as compared with isolates in compost from California, which had no resistance to this antibiotic.	[43]
	N.A./Canada	Broiler	Litter samples	9	All isolates were multiresistant to at least 7 antibiotics. Resistance to AMO, TIO, TET, and SA was the most prevalent.	[44]
Enterococcus	2006/US	3 broiler farms	Litter samples	N.A.	Resistance levels to CLI and ERY were 68%, 18%, respectively. No isolates were found to be resistant to VAN.	[45]
	N.A./US	60 chicken houses	Litter samples	N.A.	ERY-resistant bacteria were only isolated from litter samples collected from farms that had used the drug.	[41]
Providencia	N.A./US	Turkey	Fecal samples	11	Isolates were found to be resistant to TET, MAC, and SA groups.	[46]
Staphylococci	2006/US	3 broiler farms	Litter samples	N.A.	Resistance levels to CLI and ERY were 0% and 57%, respectively.	[45]
	N.A./US	Poultry	Litter samples	60	ERY-resistant bacteria were only isolated from litter samples collected from farms that had used the drug.	[41]
Streptococcus	N.A./US	Poultry	Litter samples	60	ERY-resistant bacteria were only isolated from litter samples collected from farms that had used the drug.	[41]

[a]N.A., not applicable; [b]NAL: nalidixic acid, ENR: enrofloxacin, SAR: sarafloxacin, AMP: ampicillin, TET: tetracycline, CAM: chloramphenicol, KAN: kanamycin, AMO: amoxicillin, TIO: Ceftiofur, SA: sulfonamide, CLI: clindamycin, ERY: erythromycin, VAN: vancomycin, MAC: macrolide

The gastrointestinal tracts of animals are the natural habitats for most of the enteric pathogens. After being defecated in feces, these pathogens are immediately exposed to a hostile environment with numerous microorganisms to compete for limited nutrients. Botts *et al.* [47] and Tucker [48] found that *S.* Pullorum and *S.* Gallinarum persisted much longer in fresh chicken litter than in built-up litter. Other studies have also shown that some pathogens in fresh chicken manure can initially grow to higher numbers under favorable environmental conditions. Himathongkham and Riemann [49] reported that *E. coli* O157:H7 and *L. monocytogenes* were able to multiply by as much as 100-fold for a period of 2 days in fresh chicken manure at 20 °C, whereas *S.* Typhimurium populations remained stable. Therefore, special attention should be paid to the initial disinfection processing so as to effectively eliminate pathogens from chicken litter.

When animal wastes are introduced into the agricultural field, the antagonistic effect of indigenous soil microorganisms and the hostile condition of soil microcosm are possible factors influencing the length of time that pathogens can persist [50]. According to the Standards for the Growing, Harvesting, Packing, and Holding of Produce for Human Consumption (Proposed Rule) proposed in the U.S. Food and Drug Administration (USFDA) Food Safety Modernization Act (FSMA) [51], growth of human pathogens in biological soil amendments of animal origin could result in the amendment acting as an inoculum that spreads pathogens to covered produce growing area, leading to a likelihood of produce contamination. Previous studies have reported the growth and persistence of human pathogens in chicken manure and manure-amended soil. Islam *et al*. [52] reported that *S*. Typhimurium persisted for 203 to 231 days in soils amended with poultry compost, dairy compost, and alkaline-pH-stabilized dairy compost. As pathogens most commonly associated with fresh produce outbreaks, including *E. coli*, *Salmonella*, and *Listeria*, are unlikely to survive at detectable population levels in soil after 270 days [51], it is proposed by FSMA that the waiting period (application interval) between the application of untreated biological soil amendments of animal origin (such as untreated chicken litter) and the harvest of covered produce should be 9 months provided the material is used in a manner that does not contact covered produce during application and minimizes the risk of contamination after application.

FOOD SAFETY, AND HUMAN AND ANIMAL HEALTH ISSUES ASSOCIATED WITH CHICKEN LITTER OR CHICKEN LITTER-BASED ORGANIC FERTILIZERS

Poultry is one of the commodities most commonly associated with foodborne disease outbreaks in the preceding years. During 2009–2010, the commodities in the 299 outbreaks associated with the most illnesses were eggs (27% of illnesses), beef (11%), and poultry (10%) [53]. Although foodborne disease outbreaks caused by bacterial pathogens reported so far have rarely been linked directly to chicken litter or chicken litter-based organic fertilizers, their risks to contaminate food or environment is considerably high. And there have been some food safety and human health issues associated with chicken litter in recent years.

Feeding poultry litter to dairy and beef cattle is a means of disposing of a waste product while concurrently supplying a low-cost protein feed to cattle [28]. Processed chicken litter has been used as a feed ingredient for almost 40 years in the US [15]. Cattle have the ability to digest low-cost feedstuffs, such as chicken litter, that are not suitable for other livestock species. However, from the hygienic perspective, raw chicken litter may contain some bacterial pathogens, as noted above. Salmonellosis has been reported in cattle that were fed improperly composted broiler litter [54,55]. Chicken litter was also implicated as a possible source of chronic histoplasmosis case in 2003, which was caused by inhaling fungal spores released by *Histoplasma capsulatum* when the litter was spread on a pasture [56]. Moreover, some of the fungal species that are indigenous to the manure or litter can result in the production of mycotoxins. The specifications suggested by the Association of American Feed Control Officials (AAFCO) require that processed animal waste products as feed ingredient must be free of human pathogenic microorganisms, which could be harmful to animals or could result in residues in human food products or by-products of animals at levels in excess of those allowed by State or Federal statute or regulation [15]. Hence, proper processing to reduce the amount of these microorganisms or render the waste absent of pathogens is required. In addition, feed additives such as antibiotics are also added into poultry diet, which can be excreted as waste by-products used for cattle feed. Although the use of non-therapeutic levels of antibiotics in animal feed is approved and regulated by the USFDA [57], there is still limited information about the specific types and amounts of antibiotics that should be used for this purpose.

Pathogens can be transmitted to humans directly through contact with poultry litter or indirectly through contaminated poultry products. Water may also become contaminated by runoff either from poultry facilities or from excessive land application of poultry waste [58]. Runoff can possibly carry pathogens from the original site of animal manure-applied agricultural fields to water bodies serving as irrigation, drinking, or recreational water sources [59]. Clear understanding of the transport of pathogens potentially present in poultry wastes and its runoff is essential for the establishment of effective control strategies to reduce the adverse impact on environment, food safety, and public health. Sistani *et al.* [60] compared two methods of poultry litter application, surface broadcast and subsurface banding, to investigate the influence of application methods on *E. coli* concentration in runoff from tall fescue pasture. *E. coli* concentration was found to be significantly higher in runoff from broadcast application than subsurface banding treatment. They concluded that subsurface banding of poultry litter into perennial grassland can greatly reduce pathogen losses in runoff as compared to surface-broadcast application. Therefore, the traditional surface-broadcast application of chicken litter onto agricultural land may result in high levels of pathogens on the soil surface that could be potentially transferred to runoff water.

Adequate control of pathogens may require multiple control interventions to achieve significant reduction of pathogens in a poultry waste management system. And there is a need for some good management practices to reduce potential human exposures to these pathogens by effectively controlling them during chicken litter or chicken litter-based organic fertilizer processing.

CONTROL OF PATHOGENS IN CHICKEN LITTER OR CHICKEN LITTER-BASED ORGANIC FERTILIZERS

As previously stated, a variety of pathogenic bacteria may be present in animal wastes including those destined for composting, presenting a risk of human infection when they are utilized for land application. Composting is commonly employed as a pathogen control technique to recycle animal wastes back into the soil to improve its fertility [61]. Heat treatment after composting or without composting is also recommended to reduce or eliminate potential bacterial pathogens in animal wastes.

Meanwhile, some physical, chemical, and biological methods have been developed as alternative disinfection techniques for animal waste processing.

Composting

The interest in composting has greatly increased in recent years because of the need for environmentally acceptable animal waste treatment technologies and also the demand for organic fertilizers in organic agricultural production system. Composting of animal wastes is a spontaneous bio-oxidative process to produce more uniform, concentrated, and safe final products compared to fresh manure, allowing for easy spreading in the soil and also significant elimination of pathogens [62]. Furthermore, initial capital as well as operating and maintenance costs for composting are lower compared with other treatment techniques [63]. Composting can thus be considered as an effective technique which adds value to poultry waste for agricultural applications.

Composting Process

Before land application, chicken litter is usually built-up inside the chicken house during the growing season of broilers. Build-up is a common method of storing solid animal manure or used as bedding materials until it can be composted or applied to cropland as fertilizer. Barker *et al*. [64] observed that the middle and bottom sections of the built-up broiler litter bed provided a less favorable environment for anaerobes and coliforms than the top section, as the temperature required to reduce or eliminate bacterial loads are not achieved as they are at deeper layers.

Compared to build-up, composting is a controlled process of mixing organic wastes with other ingredients in an appropriate ratio to optimize microbial growth [65]. Composting is typically the biological decomposition process of biodegradable organic wastes in a predominantly aerobic environment by a consortium of microorganisms. Generally, it is a fast biodegradation process, which takes 4–6 weeks of microbial action to break down organic materials to stable and usable organic substances called compost. Composting allows easy handling and elimination of pathogens (including human and plant pathogens), along with the volume reduction of the wastes and the destruction of weed seeds. However, disadvantages of composting are also documented, such as loss of nitrogen and other nutrients during composting, cost of installation and labor, odor, and requirement for available land for storage and operation [62,66]. The cost of transporting chicken litter is a major obstacle facing the more efficient use of this poultry by-product. After composting, the bulk density of chicken litter is increased, which can reduce the cost of transportation [3]. Since composting can result in considerable nitrogen loss, compost producers may need to add amendments, such as straw, peat, woodchip, paper waste, aluminum sulfate, and zeolite, to the litter to reduce ammonia volatilization during this process [67]. Moreover, although the objective of composting is generally to achieve a stable waste product, it may also affect both the total content and the composition of metals, such as cadmium, copper and zinc, in poultry manure by-products [68].

The process of composting is typically divided into four main phases based on temperature and active microbial community: hemophilic, thermophiles, cooling, and maturation phases [69]. Microbial activity is critical for a satisfactory composting process, in which mesophilic, thermo tolerant, and thermophiles bacteria, actinomycetes, and fungi are all extensively involved [70]. Aerobic microbial decomposition generates sufficient heat to increase the temperature of compost mixtures to the thermophiles zone (45 to 75 °C). Temperatures reached in a well-managed compost operation should be within a range of 55 to 65 °C [71]. Such temperatures are well above the thermal death points of hemophilic pathogens, such as *E. coli* O157:H7 and *Salmonella* spp. [72]. Besides high temperature, other mechanisms are also known to get involved in the inactivation of foodborne pathogens during composting, including microbial antagonism, production of organic acids, pH change, desiccation and starvation stresses, exposure to ammonia emission, and competition for nutrients [1].

Currently, within the United States, composting of animal wastes is not regulated by any federal agencies. The National Organic Program, administered by the U.S. Department of Agriculture (USDA), includes a composting standard in 7 CFR 205.2 that is aimed to maximize soil fertility and is required to achieve —USDA Certified Organic□ status [73]. Similar to USEPA specifications described in 40 CFR Part 503 for regulating land application of Class A composted sewage sludge [74], the FSMA [51] proposed standards for two specific scientifically valid controlled composting processes: (1) Static composting that maintains aerobic conditions at a minimum of 55 °C for three days and is followed by adequate curing, which includes proper insulation; and (2) turned windrow composting to maintain aerobic conditions at a minimum of 55 °C for 15 days with a minimum of five turnings, and is followed by adequate curing, which includes proper insulation. However, a slightly different temperature and time criterion is recommended by the guidelines for composting dead poultry proposed by USDA [65] and adapted from those developed by McCaskey [75], which requires that when the compost has achieved a temperature greater than 50 °C for at least five days, the composting process is adequate to eliminate *L. monocytogenes*, *E. coli* O157:H7, and *S.* Typhimurium.

Composting has been proved to be an effective method to produce organic fertilizers in order to treat the ever-increasing volume of poultry wastes [76,77], which converted the soluble nutrients to more stable organic forms, thus increasing their bioavailability while reducing their susceptibility to loss when applied to agricultural land. Several studies have demonstrated that composting can be an effective way of lowering the level of foodborne pathogens in chicken litter. In some cases, moreover, common foodborne pathogens can be completely eliminated during the composting process. Martin *et al.* [29] reported that no *E. coli* O157:H7 and *Salmonella* spp. was detected in 64 composted poultry litter samples. Brodie *et al.* [78] also observed the complete elimination of *Salmonella, C. jejuni*, and *L. monocytogenes* from poultry compost when temperature exceeded 55 °C. In the work of Macklin *et al.* [79], chicken litter samples inoculated with *Salmonella* and *C. perfringens* were collected after seven days to determine the population of inoculated bacteria that survived. These pathogens were completely eliminated from the composted samples, while it was still recoverable from the samples that were uncomposted. Results from the study of Silva *et al.* [80] also indicated that the final compost of poultry manure was free of fecal coliforms and *Salmonella* spp., although a thermophiles phase (temperature > 40 °C) was not verified in the compost pile. Additionally, the findings of Guan *et al.* [81] demonstrated that composting of chicken manure, when managed to produce sufficiently high temperature, could reduce or degrade heat-sensitive genetically modified *Pseudomonas chlororaphis* and their transgenes. Therefore, available information on composting of poultry wastes indicates that composting is a suitable and environmentally sound method of reducing or eliminating foodborne pathogens. Furthermore, it should be emphasized that proper composting management is needed to ensure that the process achieves the target level of the time–temperature combination for killing pathogens.

Pathogen Persistence and Regrowth after Composting

The die-off of pathogens during composting may not be extensive or uniform throughout the composting heaps or piles and depends on some environmental factors, such as microbe type, manure physic-chemical characteristics, aeration, and temperature pattern [62]. Therefore, in some cases, a traditional composting technique may not always be sufficient to ensure complete inactivation of pathogens within the entire compost mass. Consequently, persistence of pathogens in poultry compost has been reported. The surface of fresh compost has been identified as the critical location for pathogens to extend the survival or serves as the source of cross-contamination with the rest of the compost mass during heap turning or with the ambient environment. Shepherd *et al.* [30] detected that 26% and 6.1% of the surface and internal samples from poultry compost heaps were positive for *Salmonella* during the first phase of composting, respectively. Their results indicated that the conditions at the compost surface are suitable for pathogen survival, and that the complete composting process, including both heating and curing phases, should be confirmed before the compost is considered a finished, pathogen-free product. In another study, compost temperature of 55 °C was unable to inactivate *Salmonella*, *E. coli* O157:H7 and *L.* monocytogenes for more than eight days in poultry compost [82]. Macklin *et al.* [79] also recovered *C. perfringens* from five out of six interior chicken litter samples after seven days during in-house composting. Erickson *et al.* [83] conducted a field study on the fate of three avirulent pathogen surrogates (gfp-labeled *E. coli* O157:H7 and *L. innocua* and avirulent *S.* Typhimurium) in static composting piles of chicken litter and peanut hulls. *Salmonella* was detected by enrichment in sub-surface samples of static composting piles up to 14 days. Indicator microorganisms were only detected by enrichment in surface samples during the summer after four days of composting, while *E. coli* O157:H7 and *L. innocua* were still detectable by direct plating after 28 days in compost piles during the fall and winter trials. All three bacteria could be detected by enrichment in surface samples for 56 days of composting during the winter.

Besides the survival of pathogen during composting, there is also a concern over the possibility of regrowth due to the outside recontamination under open air composting environments and during storage as well. In the study of Kim *et al.* [10], *E. coli* O157:H7 increased from *ca.* 1 to 4.85 log CFU/g in autoclaved dairy compost after seven days, suggesting that a small portion of pathogenic cells that survive the composting process or are cross-contaminated from the environment could multiply to high populations under favorable conditions. In addition, studies have also demonstrated that pathogen growth in compost is influenced by several environmental variables, such as moisture content, temperature, background microflora, and nutrient availability of the composted solids [10,77]. To ensure the microbiological safety of composted chicken litter, environmental factors supporting potential pathogen regrowth after composting need to be identified and monitored. Additionally, outdoor composting is generally exposed to fluctuating environmental conditions, animal intrusion, and reduced efficiency of composting due to climatic conditions, and is not homogeneous in nature and prone to having —cold-spots□ that are not properly treated, even with complete turning [84]. It is thus possible that composting can result in a treated chicken litter that may continue to harbor a low level of human pathogens of food safety concern.

Composting Stress and Stress-Induced Cross-Protection

Composting is a very complex and dynamic biological process, which may pose a significant challenge and create many hostile stresses for the survival and growth of human pathogens. Bacterial stress is generally defined as a physical, chemical, or nutritional condition insufficiently severe to kill but leading to sublethal injury of microbes [85]. Some common types of stress associated with the composting process include desiccation, heat shock, and acid stresses [11,12,86].

1. Desiccation stress. During composting, moisture level in the compost mixture, especially at the surface of compost pile, is reduced rapidly due to evaporation and the self-heating during the thermophiles phase [9]. Water loss through the

desiccation process is an important factor affecting the survival and persistence of bacterial pathogens in low-water-activity environmental habitats, such as soil, sand, and compost surface [86].

2. Heat shock stress. Heat shock occurs when microorganisms are exposed to temperatures above their normal growth range [87]. Temperature during composting process increases gradually, from ambient temperature to the hemophilic range and then to the thermophiles phase, which may consequently cause heat shock or stimulate a concomitant genetic and physiological heat shock response in some population of pathogenic bacteria [86]. Especially during the extended hemophilic phase of composting, some bacterial cells may become acclimatized to sub lethal high temperatures before lethal temperatures are reached, allowing them to survive and, in some cases, multiply under stressful conditions. In support of this notion, results of Singh et al. [11] revealed that heat-shocked E. coli O157:H7, Salmonella, and L. monocytogenes at 47.5 °C survived longer in dairy compost than non-heat-shocked cells at composting temperatures of 50, 55 and 60 °C.
3. Acid stress. Acid stress can occur in low pH conditions when H+ ions cross the bacterial cell membrane and create an acidic intracellular environment. Acid resistance is especially crucial for foodborne pathogens that must survive the hostile acidic condition in the stomach before entering and colonizing the small intestines or colon [88]. Pathogenic cells present in compost of animal origin may become acid-adapted as they are exposed to acidic condition when passing through the gastric tract.

The presence of stressed microorganisms in manure and compost could pose significant public health concerns when applying the contaminated compost to arable land. Stressed cells in chicken litter or chicken litter-based organic fertilizers can initially go undetected during routine microbiological analysis; however, subsequent resuscitation under suitable environments may allow for significant growth and also possible production of toxins and other virulence factors [89]. Many stressed pathogens either retain or exhibit enhanced virulence and invasion, thus making their detection crucial to ensure food safety [90,91]. *S.* Enteritis's PT4 with enhanced heat and acid resistance has been reported to be more virulent in mice and more invasive in chickens than the non-resistant reference strain [92]. Interestingly, some stresses that are originally part of the host's defense system are very similar to those occurred in the natural environment [93]. Therefore, it is reasonable to speculate that pathogens may treat stresses encountered during poultry waste composting or build-up as a signal for the expression of virulence factors. When pathogen-contaminated poultry wastes or their composted products are used as organic fertilizer and soil amendment, pathogens with increased virulence could transfer to produce in the field and thus cause serious foodborne disease outbreaks. However, further research is needed to verify this hypothesis.

Bacteria typically respond to stresses by altering their cellular morphology, membrane composition, biological metabolism, and virulence. Such stressed microorganisms produce a series of stress responses that can afford cross-protection against other stresses, indicating that the adaptation to a single sublethal stress may also enhance the tolerance to multiple lethal stresses. In fact, bacteria, especially foodborne pathogens, are frequently exposed to environmental stresses that cross-protect them against various other stresses [85]. Bacterial cells can gradually adapt to the hostile sublethal conditions, causing an adaptive response accompanied by a temporary physiological change that may result in an enhanced stress tolerance [94]. The general stress response identified in most Gram-negative bacteria, such as *E. coli*, *S.* Typhimurium, and *P. aeruginosa*, is regulated by the sigma factor, RpoS (σ^S) [95]. Induction of RpoS makes bacteria more resistant to environmental stresses, such as high and low temperatures, prolonged starvation, osmotic shock, pH stress, and oxidative stress [96]. Bacteria defective in the gene (*rpoS*) for RpoS synthesis have proved to be more sensitive to different adverse conditions [97]. Van Hoek *et al.* [98] reported that a fully functional RpoS system can provide an advantage for survival in the manure-amended soil environment. In their study, *E. coli* O157 isolates capable of long-term survival in manure-amended soil were all absent of mutations in their *rpoS* gene; however, the strains not capable of long-term persistence were mutant in their *rpoS* gene.

Other Treatments

The common practice of applying poultry wastes to soil as a source of nutrients to crops is of paramount importance in sustainable agriculture. While composting, to some extent, is an effective method for reducing pathogen concentrations in animal manure, pathogens can still survive in the composted products and contribute to soil contamination. Additional appropriate processing and control measures should therefore be adopted to minimize the pathogen contamination of chicken litter. The FSMA also suggests that such biological soil amendments with a residual population of pathogens after composting should require a multiple hurdle approach to minimize the likelihood of introducing pathogens to the field upon which they are applied [51]. There are a variety of physical, chemical, and biological approaches that have the potential to effectively disinfect chicken litter. Several treatment processes have been designed to operate at conditions capable of disinfecting bacterial pathogens in chicken litter, and the extent to which they actually reduce these microorganisms have also been characterized in laboratory and pilot scale field studies.

Physical Treatment Techniques

Physically heat-drying manure or compost to low moisture content after composting or without composting can reduce the volume and weight, which thus lowers transportation costs, though it requires significant energy inputs. Heat-dried products can be much easier to handle and spread uniformly over an agricultural field, especially after they have been further processed into pellets. Table 3 presents the temperature–time requirements and acceptance criteria for the physical heat-drying of soil amendments.

Although a number of organizations or federal agencies offer independent protocol to ensure safe and effective heat treatment for animal manure or biosolids, there were no defined heat sources (dry *vs.* moist heat), varying time–temperature requirement, and microbial standards among groups. Research on the inactivation effects of thermal processing on pathogens in chicken litter has yielded generally satisfactory results. The main factors influencing pathogen reductions in animal manure by these processes are temperature, duration of treatment, and moisture level [2]. Wilkinson *et al.* [1] could not detect any *S.* Typhimurium in fresh poultry litter (30%, 50% and 65% moisture contents) after 1 h wet heat treatment at 55 or 65 °C in water bath. Kim *et al.* [2] found that a 7-log reduction of *Salmonella* spp. in fresh chicken litter (30%, 40% and 50% moisture contents) could be achieved by dry heat treatment at 80 °C in a conventional oven for 44.1 to 63.0 min. When they investigated the effect of chicken litter freshness on heat resistance profiles of *Salmonella, Salmonella* cells in aged chicken litter survived significantly longer than those in fresh chicken litter under any conditions. Ghaly and Alhattab [99] reported that the drying process at 40, 50 and 60 °C reduced the populations of bacteria, yeast and mold, and *E. coli* in chicken manure by 65.6%–99.8%, 74.1%–99.6% and 99.9%, respectively. Salmonellae were detected in the raw manure and the dried manure samples collected from the 3 cm deep manure layer after drying at 40 °C but not at 50 and 60 °C. Their results indicated that the higher the drying temperature and/or the thinner the manure layer, the more destruction of microorganisms in the dried manure. It should be noted that the aforementioned results on temperature–time combination requirements for eliminating *Salmonella* varied among studies. However, comparisons between these different studies should be conducted with precaution due to the differences in the composition and moisture level of chicken litter material, *Salmonella* strain, and also heating source. For example, Messer *et al.* [100] found that four different kinds of bacterial pathogens in chicken litter were destroyed by dry heat at different temperatures and within different times. *Arizona* spp., *S.* Pullorum, *S.* Typhimurium, and *E. coli* were destroyed by heat at 47.2 °C for 30 min, 62.8 °C for 30 min, 62.8 °C for 60 min, 68.3 °C for 30 min, respectively. Also, the physiological stage may contribute to the difference in heat resistance of pathogens in chicken litter. The work by Chen *et al.* [101] demonstrated that *Salmonella* cells adapted under a desiccation condition survived much longer in aged chicken litter as compared to non-adapted cells when exposed to the same dry heat treatment at 70, 75, 80, 85 and 150 °C. An obvious variability in heat resistance profile among *Salmonella* serotypes was also observed during thermal exposure, since *S.* Senftenberg and *S.* Typhimurium exhibited higher levels of heat resistance than *S.* Enteritidis and *S.* Heidelberg.

Table 3. Temperature–time requirements and acceptance criteria for the physical heat-drying of soil amendments.

Source	Soil amendment	Temperature-time requirement	Acceptance criterion	
			Moisture level	Microbial level
USEPA [74]	Biosolids	Either the temperature of the biosolids >80 °C or the wet bulb temperature of the gas in contact with the biosolids as the biosolids leave the dryer >80 °C	<10%	For Class A biosolids, fecal coliforms: <1000 MPN/g dry weight or *Salmonella*: <3 MPN/4 g dry weight
National Organic Program [102]	Animal manure	>65 °C for >60 min	<12%	Fecal coliforms, *Salmonella*, and *E. coli* O157:H7: negative
European Union [103]	Animal manure	>70 °C for >60 min	N.A.	*E. coli* or *Enterococaceae*: <1000 MPN/g *Salmonella*: absence in 25 g of sample
California Leafy Green Products Handler Marketing Agreement [104]	Animal manure	Either the process has been validated by a recognized authority or is subject to 150 °C for 60 min	<30%	Fecal coliforms, *Salmonella*, and *E. coli* O157:H7: negative or less than detection limit

According to the FSMA [51], chicken manure may be physically heat-treated to create a dried, pelleted material that is functionally sterile due to the high heat used during production; however, it has been observed that if the heat treatment is not uniform, the end product may still harbor human pathogens and pose a likelihood of the material being recolonized by these pathogens, leading to the possible contamination of any covered produce to which it is applied. Moreover, physically heat-treated poultry manure pellets would be expected to have limited microorganism content including competitive native microflora of foodborne pathogens, which may provide an opportunity for the potential growth of foodborne pathogens in the event of pellet contamination. The significant energy costs for heat-drying manure or compost at high temperature are also in great contrast to the self-heating generated by microbial respiration during the composting process. The physical heat-drying of manure or compost may largely increase the volatilization of ammonia–nitrogen and reduce the total nitrogen level in the finished products. Additionally, manure or compost that is physically heat-dried at a high temperature instead of going through a curing phase at an ambient temperature is not considered to be as microbiologically active as the composted products.

Besides traditional thermal processing, some other non-thermal physical methods have also been studied to control foodborne pathogens in chicken litter. Barbour *et al.* [105] evaluated the impact of soil polarization on reduction of indicator microorganisms of fecal contamination in chicken manure-treated soils. The percentage reductions in counts in chicken manure-treated soil collected at 20 cm depth after solarization were: 26.3% (*S. aureus*), 45.5% (total bacteria), 71.3% (fungi), 81.8% (*C. perfringes*), 92.6% (fecal coliforms) and 100.0% (non-lactose fermenting bacteria). Significantly, UV disinfection may be an effective treatment for reducing pathogen concentrations in animal wastes, while retaining the nutrient content for crop production. Oni *et al.* [106] investigated the effect of UV radiation on the survival of *Salmonella* when present on dried turkey manure particles. They observed that *Salmonella* exposed to UV in a thin layer of cells in saline that were directly placed in a petri dish showed a 5-log decline within 80 min, as compared to the 1.5-log decrease in turkey manure dispersed as a thin layer on a petri dish, suggesting the presence of manure particles significantly ($p < 0.05$) protected *Salmonella* from UV exposure.

Chemical Treatment Techniques

Alum treatment has been widely used to reduce pathogens before land application of chicken litter [107], and it can also be applied as an effective way to reduce ammonia volatilization and water-soluble phosphorus runoff from poultry litter in chicken houses [108]. Pathogens in poultry wastes, such as *Salmonella* and *Campylobacter*, have been shown to be reduced significantly or eliminated by alum treatment [109–111]. Rothrock *et al.* [107] used denatured gradient gel electrophoresis (DGGE) and quantitative real-time polymerase chain reaction (PCR) to characterize pathogenic microbial communities in alum-treated poultry litter. Alum addition (10% wt/wt) resulted in significant reductions in *C. jejuni, E. coli,* and *Clostridium/Eubacterium* populations by the end of the first month. The concentrations of *Salmonella* spp. were below detection limit ($<5 \times 10^3$ cell/g litter) throughout the entire experiment. Similarly, when Gandhapudi *et al.* [108] studied the immediate effects of alum on potential nitrification in poultry manure, slurries with alum-treated poultry manure reduced the population of fecal bacteria, presumably because of the pH stress that suppressed bacterial growth. However, labor requirements to apply alum to chicken litter and mix them sufficiently may be a limiting factor.

Lime in the form of quicklime (CaO) or hydrated lime [$Ca(OH)_2$] has also been reported to effectively disinfect and stabilize chicken litter by raising the pH of the waste to 12 or increasing the temperature by exothermic reaction, which are beyond the tolerance ranges of most enteric pathogens [112,113]. The disinfection efficacy of quicklime is also attributed to the dehydration in the poultry litter. With reduced water availability, bacteria will need more energy to absorb water from the litter for metabolic processes, making survival more difficult or even impossible [114]. However, there may be some additional costs to consider such as human labor to mix and haul the lime. Maguire *et al.* [115] applied 10% quicklime to 20% solid broiler litter and reduced the total plate counts from 793,000 to 6500 CFU/mL, which, as they indicated, could reflect the fate of many organisms, such as *Salmonella*. Bennett *et al.* [116] found that the addition of lime (5%, 10% and 20%) to poultry litter significantly reduced the recovery incidence of *S.* Enteritidis within 24 h. Nevertheless, Bennett *et al.* [117] verified that the effects of 0%, 1%, 2% and 5% of lime addition on turkey litter did not lead to a reduction in *Campylobacter* or *Salmonella,* which are contradictory to the above findings, but lime treatment did reduce the population of aerobic bacteria. This difference may be explained by differences in the composition of poultry litter sample and also the lime concentration applied.

Ammonia can also cause a significant reduction of non-spore-forming pathogens in a stacked manure pile [118,119]. There is, however, a lack of comprehensive research concerning this area. The limited information available on the effect of ammonia gas on common pathogens in chicken litter shows that generally it results in a significant reduction of pathogens. Himathongkham and Riemann [49] evaluated the effect of drying and/or exposure to ammonia on *S.* Typhimurium, *E. coli* O157:H7 and *L. monocytogenes* in chicken manure. Drying to a moisture content of 10% followed by gassing with ammonia for 72 h as in an amount of 1% of the manure wet weight resulted in an 8-log reduction of *S.* Typhimurium and *E. coli*

O157:H7, and 4-log reduction for *L. monocytogenes*. However, some environmental issues associated with potential pollution impacts of applying ammonia to chicken litter include ammonia emissions as aerial contaminants [120].

Fontenot *et al.* [121] investigated the effects of various processing techniques on sterilizing broiler litter. Heating at 150 °C for 1 or 2 h, dry heating at 100 °C for 48 h, autoclaving, beta-propiolactone treatment, or ethylene oxide fumigation was ineffective in completely sterilizing the litter. The only effective method was dry heat at 150 °C for a minimum of 3 h. Extensive studies were conducted by Caswell *et al.* [122] investigating the effectiveness of paraformaldehyde addition and ethylene oxide fumigation on pasteurization of broiler litter. They found that ethylene oxide treatment and dry heat at 150 °C following the addition of paraformaldehyde were all effective in reducing coliforms and total bacteria in broiler litter from >30,000 CFU/g to <2000 CFU/g. Seltzer *et al.* [123] found that the addition of 1, 3, or 7 g of paraformaldehyde per 100 g of fresh chicken feces reduced total bacteria counts from 2.2×10^9 CFU/g for untreated feces to 1.6×10^8, 1000 and 0 CFU/g, respectively.

Chlorine is an effective disinfection method commonly used for drinking water; however, the high organic matter found in chicken litter may largely diminish the effectiveness of chlorine. Murray *et al.* [124] concluded that chlorination, while initially reducing the total population of bacteria in sewage, may greatly increase the proportions of antibiotic-resistant pathogenic bacteria. It is still uncertain whether chlorination specifically induces modifications and changes in antibiotic resistance in some bacterial populations. Additionally, Munir *et al.* [125] reported that chlorination did not contribute to a significant reduction of antibiotic resistant genes and antibiotic resistant bacteria. The efficacy of chlorination can be substantially impacted by the presence of suspended organic matter, which may significantly lower the effective chlorine concentration which bacterial pathogens can be exposed to, and therefore require a higher amount of chlorine [14]. Without extensive pre-treatment, it is possible that manure-derived waste will have substantial interfering substances that will make chlorination less effective. Furthermore, the chemical reactions between chlorine and organic matter, nitrogen-containing compounds, or ammonia in chicken litter or chicken litter-based organic fertilizers when they are exposed to each other may also generate carcinogenic by-products, posing great risk to human health when applying them to agricultural land [126].

While effective in reducing pathogen levels, most of the current chemical treatment techniques are not economically feasible for large-scale chicken litter processing. Meanwhile, the interferences of high loads of endogenous non-pathogenic microorganisms in chicken litter may also influence the efficacy of chemical treatments on the killing of target pathogens.

Biological Control Techniques

To ensure that pathogens are eliminated from chicken litter or chicken litter-based organic fertilizers before application to agricultural land, some innovative biological control approaches have also been explored. The results obtained in the study of Erickson *et al.* [127] revealed the significant inactivation effect of soldier fly larvae (*Hermetia illucens* L.) against foodborne pathogens in chicken manure. After two days, *Salmonella* population in chicken manure containing larvae had decreased by 4-log CFU/g. A 6-log reduction in the population of *E. coli* O157:H7 in chicken manure was also observed after adding larvae for three days. Yongabi *et al.* [128] designed a simple plastic anaerobic digester to disinfect contaminated poultry feces while providing biogas and pathogen-free fertilizer. Following anaerobic digestion of poultry feces for 37 days, both coliform and *E. coli* counts decreased drastically. It was also reported by Krylova *et al.* [129] that high levels of ammonium (>30 g/L) during anaerobic digestion of poultry litter resulted in a decrease in the numbers of all physiological microbial groups. A study on bacteriophage has also demonstrated its effectiveness for biological control of pathogens in compost. Heringa *et al.* [130] applied a five-strain bacteriophage mixture to dairy manure compost inoculated with *S.* Typhimurium. They found that bacteriophage treatment resulted in a greater than 2-log-unit reduction within 4 h. And since bacteriophages are host-specific, unlike physical or chemical treatments, it will not significantly alter the background microbial community in animal wastes. Therefore, application of bacteriophage to chicken litter or chicken litter-based organic fertilizers may be another promising biological control technique during the preharvest stage.

CONCLUSIONS

Raw chicken litter has been widely applied to arable land as organic fertilizer or soil amendment to improve the soil fertility and structure. To prevent possible microbiological safety issues for the environment and food crops grown in the field, practical and effective treatments should be developed specifically for raw chicken litter prior to land application. Composting, commonly used on farms, can inactivate large populations of human pathogens; however, studies have revealed that some pathogens can survive the composting process due to improper composting or cross-contamination. As a result, a small population of pathogenic cells may survive or regrow in the finished compost products under favorable conditions. Physical, chemical, and biological treatments can be alternative ways for pathogen inactivation but may not always lead to the complete elimination of foodborne pathogens in chicken litter or chicken litter-based organic fertilizers. Furthermore, some cells may become stress-adapted during build-up or composting, which cross-protect them against these subsequent treatments. Based on the hurdle concept, each kind of treatment can be used in combination with other disinfection strategies to potentiate microbial lethality. In order to effectively inactivate pathogens in chicken litter, it would be plausible to design a multi-step treatment with composting as the first step to kill large populations of pathogens, and then apply additional treatments to further eliminate the remaining cells. These systems with multiple treatments involved can be efficient in eliminating pathogens in chicken litter when proper control measures are in place and adopted.

Although chicken litter is considered a potential source of foodborne pathogens, this does not suggest that every portion of the litter contains all the various kinds of pathogens that have been reported or that they will be present at maximally reported prevalence. Nonetheless, treatment techniques should still be developed to inactivate the most resistant and persistent types of pathogens possibly to be encountered. Most of the studies on physical, chemical, and biological treatment techniques have attempted to quantify reductions of different bacterial pathogens or indigenous microorganisms in chicken litter or its composted products. Some estimates of pathogen reductions are uncertain and based only on limited lab studies with few pathogens, including indicator microbes (primarily fecal coliforms). However, it is still not clear whether the fate of such fecal indicator bacteria properly represents the responses of various human pathogens. Moreover, not all fecal coliforms or tested pathogens arise from animal feces, and they have some non-fecal environmental sources, which makes it more difficult to investigate the fate of pathogens in animal wastes during different treatments. Therefore, future studies should focus on evaluating pathogen survival for different treatments using a wide range of conditions commonly encountered during build-up or composting.

ACKNOWLEDGEMENT

This work was partially supported by a grant from the Center for Produce Safety, University of California, Davis.

Conflicts of Interest

The authors declare no conflict of interest.

REFERENCES

1. Wilkinson, K.G.; Tee, E.; Tomkins, R.B.; Hepworth, G.; Premier, R. Effect of heating and aging of poultry litter on the persistence of enteric bacteria. *Poult. Sci.* 2011, *90*, 10–18.
2. Kim, J.; Diao, J.; Shepherd, M.W., Jr.; Singh, R.; Heringa, S.D.; Gong, C.; Jiang, X. Validating thermal inactivation of *Salmonella* spp. in fresh and aged chicken litter. *Appl. Environ. Microbiol.*
3. 2012, *78*, 1302–1307.
4. Moore, P.A., Jr.; Daniel, T.C.; Sharpley, A.N.; Wood, C.W. Poultry manure management: Environmentally sound options. *J. Soil Water Conserv.* 1995, *50*, 321–327.
5. Enticknap, J.J.; Nonogaki, H.; Place, A.R.; Hill, R.T. Microbial diversity associated with odor modification for production of fertilizers from chicken litter. *Appl. Environ. Microbiol.* 2006, *72*, 4105–4114.
6. Wilkinson, S.R. Plant nutrient and economic value of animal manures. *J. Anim. Sci.* 1979, *48*, 121–133.
7. Chinivasagam, H.N.; Redding, M.; Runge, G.; Blackall, P.J. Presence and incidence of foodborne pathogens in Australian chicken litter. *Br. Poult. Sci.*

2010, *51*, 311–318.

8. Lemunier, M.; Francou, C.; Rousseaux, S.; Houot, S.; Dantigny, P.; Piveteau, P.; Guzzo, J. Long-Term survival of pathogenic and sanitation indicator bacteria in experimental biowaste composts. *Appl. Environ. Microbiol.* 2005, *71*, 5779–5786.
9. You, Y.; Rankin, S.C.; Aceto, H.W.; Benson, C.E.; Toth, J.D.; Dou, Z. Survival of *Salmonella enterica* serovar Newport in manure and manure-amended soils. *Appl. Environ. Microbiol.* 2006, *72*, 5777–5783.
10. Shepherd, M.W.; Liang, P.; Jiang, X.; Doyle, M.P.; Erickson, M.C. Fate of *Escherichia coli* O157:H7 during on-farm dairy manure-based composting. *J. Food Protect.* 2007, *70*, 2708–2716.
11. Kim, J.; Luo, F.; Jiang, X. Factors impacting the regrowth of *Escherichia coli* O157:H7 in dairy manure compost. *J. Food Protect.* 2009, *72*, 1576–1584.
12. Singh, R.; Jiang, X.; Luo, F. Thermal inactivation of heat-shocked *Escherichia coli* O157:H7,
13. *Salmonella*, and *Listeria monocytogenes* in dairy compost. *J. Food Protect.* 2010, *73*, 1633–1640.
14. Singh, R.; Jiang, X. Thermal inactivation of acid-adapted *Escherichia coli* O157:H7 in dairy compost. *Foodborne Pathog. Dis.* 2012, *9*, 741–748.
15. Doyle, M.P.; Erickson, M.C. Summer meeting 2007—The problems with fresh produce: An overview. *J. Appl. Microbiol.* 2008, *105*, 317–330.
16. Kelleher, B.P.; Leahy, J.J.; Henihan, A.M.; O'Dwyer, T.F.; Sutton, D.; Leahy, M.J. Advances in poultry litter disposal technology—A review. *Bioresource Technol.* 2002, *83*, 27–36.
17. Bolan, N.S.; Szogi, A.A.; Chuasavathi, T.; Seshadri, B.; Rothrock, M.J., Jr.; Panneerselvam, P. Uses and management of poultry litter. *World's Poult. Sci. J.* 2010, *66*, 673–698.
18. Alexander, D.C.; Carrière, J.A.; McKay, K.A. Bacteriological studies of poultry litter fed to livestock. *Can. Vet. J.* 1968, *9*, 127–131.
19. Lovett, J.; Messer, J.W.; Read, R.B. The microflora of Southern Ohio poultry litter. *Poult. Sci.* 1971, *50*, 746–751.
20. Lu, J.; Sanchez, S.; Hofacre, C.; Maurer, J.J.; Harmon, B.G.; Lee, M.G. Evaluation of broiler litter with reference to the microbial composition as assessed by using 16S rRNA and functional gene markers. *Appl. Environ. Microbiol.* 2003, *69*, 901–908.
21. Stern, N.J.; Robach, M.C. Enumeration of *Campylobacter* spp. in broiler feces and in corresponding processes carcasses. *J. Food Protect.* 2003, *66*, 1557–1563.
22. Ngodigha, E.M.; Owen, O.J. Evaluation of the bacteriological characteristics of poultry litter as feedstuff for cattle. *Sci. Res. Essays* 2009, *4*, 188–190.
23. Higgins, R.; Malo, R.; René-Roberge, E.; Gauthier, R. Studies on the dissemination of *Salmonella* in nine broiler-chicken flocks. *Avian Dis.* 1982, *26*, 26–33.
24. Renwick, S.A.; Irwin, R.J.; Clarke, R.C.; McNab, W.B.; Poppe, C.; McEwen,

S.A. Epidemiological associations between characteristics of registered broiler chicken flocks in Canada and the *Salmonella* culture status of floor litter and drinking water. *Can. Vet. J.* 1992, *33*, 449–458.

25. Li, X.; Payne, J.B.; Santos, F.B.; Levine, J.F.; Anderson, K.E.; Sheldon, B.W. *Salmonella* populations and prevalence in layer feces from commercial high-rise houses and characterization of the *Salmonella* isolates by serotyping, antibiotic resistance analysis, and pulsed field gel electrophoresis. *Poult. Sci.* 2007, *86*, 591–597.
26. Centers for Disease Control and Prevention (CDC). Surveillance for foodborne disease outbreaks—United States, 1998–2008. *Morbid. Mortal. Weekly Rep. (MMWR)* 2013, *62*, 1–34.
27. Forsythe, R.H.; Ross, W.J.; Ayres, J.C. *Salmonella* recovery following gastro-intestinal and ovarian inoculation in the domestic fowl. *Poult. Sci.* 1967, *46*, 849–855.
28. Foley, S.L.; Nayak, R.; Hanning, I.B.; Johnson, T.J.; Han, J.; Ricke, S.C. Population dynamics of *Salmonella enterica* serotypes in commercial egg and poultry production. *Appl. Environ.*
29. *Microbiol.* 2011, *77*, 4273–4279.
30. Schoeni, J.L.; Glass, K.A.; McDermott, J.L.; Wong, A.C.L. Growth and penetration of *Salmonella* enteritidis, *Salmonella* heidelberg and *Salmonella* typhimurium in eggs. *Int. J. Food Microbiol.* 1995, *24*, 385–396.
31. Jeffrey, J.S.; Kirk, J.H.; Atwill, E.R.; Cullor, J.S. Research notes: Prevalence of selected microbial pathogens in processed poultry waste used as dairy cattle feed. *Poult. Sci.* 1998, *77*, 808–811.
32. Martin, S.A.; McCann, M.A.; Waltman, W.D. Microbiological survey of Georgia poultry litter. *J. Appl. Poult. Res.* 1998, *7*, 90–98.
33. Shepherd, M.W.; Liang, P.; Jiang, X.; Doyle, M.P.; Erickson, M.C. Microbiological analysis of composts produced on South Carolina poultry farms. *J. Appl. Microbiol.* 2010, *108*, 2067–2076.
34. Smyser, C.F.; Snoeyenbos, G.H. Evaluation of several methods of isolating salmonellae from poultry litter and animal feedstuffs. *Avian Dis.* 1969, *13*, 134–141.
35. Bhargava, K.K.; O'Neil, J.B.; Prior, M.G.; Dunkelgod, K.E. Incidence of *Salmonella* contamination in broiler chickens in Saskatchewan. *Can. J. Comp. Med.* 1983, *47*, 27–32.
36. Long, J.R.; DeWitt, W.F.; Ruet, J.L. Studies on *Salmonella* from floor litter of 60 broiler chicken houses in Nova Scotia. *Can. Vet. J.* 1980, *21*, 91–94.
37. Orji, M.U.; Onuigbo, H.C.; Mbata, T.I. Isolation of *Salmonella* from poultry droppings and other environmental sources in Awka, Nigeria. *Int. J. Infect. Dis.* 2005, *9*, 86–89.
38. Nógrády, N.; Kardos, G.; Bistyák, A.; Turcsányi, I.; Mészáros, J.; Galántai, Z.; Juhász, A.;
39. Samu, P.; Kaszanyitzky, J.E.; Pászti, J.; *et al.* Prevalence and characterization

of *Salmonella infantis* isolates originating from different points of the broiler chicken-human food chain in Hungary. *Int. J. Food Microbiol.* 2008, *127*, 162–167.

40. Alali, W.Q.; Thakur, S.; Berghaus, R.D.; Martin, M.P.; Gebreyes, W.A. Prevalence and distribution of *Salmonella* in organic and conventional broiler poultry farms. *Foodborne Pathog. Dis.* 2010, *7*, 1363–1371.
41. Sidh, J.P.S.; Toze, S.G. Human pathogens and their indicators in biosolids: A literature review. *Environ. Int.* 2009, *35*, 187–201.
42. Rensing, C.; Newby, D.T.; Pepper, I.L. The role of selective pressure and selfish DNA in horizontal gene transfer and soil microbial community adaptation. *Soil Biol. Biochem.* 2002, *34*, 285–296.
43. Nandi, S.; Maurer, J.J.; Hofacre, C.; Summers, A.O. Gram-positive bacteria are a major reservoir of Class 1 antibiotic resistance integrons in poultry litter. *Proc. Natl. Acad. Sci. USA* 2004, *101*, 7118–7122.
44. Levy, S.B. *The Antibiotic Paradox: How Miracle Drugs are Destroying the Miracle*; Plenum Publishing: New York, NY, USA, 1992.
45. Khan, A.A.; Nawaz, M.S.; Khan, S.A.; Steele, R. Detection and characterization of erythromycin-resistant methylase genes in Gram-positive bacteria isolated from poultry litter. *Appl. Microbiol. Biotechnol.* 2002, *59*, 377–381.
46. Hofacre, C.L.; Cotret, A.R.; Maurer, J.J.; Garritty, A.; Thayer, S.G. Presence of fluoroquinolone-resistant coliforms in poultry litter. *Avian Dis.* 2000, *44*, 963–967.
47. Heringa, S.; Kim, J.; Shepherd, M.W.; Singh, R.; Jiang, X. The presence of antibiotic resistance and integrons in *Escherichia coli* isolated from compost. *Foodborne Pathog. Dis.* 2010, *7*, 1297–1304.
48. Furtula, V.; Farrell, E.G.; Diarrassouba, F.; Rempel, H.; Pritchard, J.; Diarra, M.S. Veterinary pharmaceuticals and antibiotic resistance of *Escherichia coli* isolates in poultry litter from commercial farms and controlled feeding trials. *Poult. Sci.* 2010, *89*, 180–188.
49. Grahama, J.P.; Evans, S.L.; Price, L.B.; Silbergeld, E.K. Fate of antimicrobial-resistant enterococci and staphylococci and resistance determinants in stored poultry litter. *Environ. Res.* 2009, *109*, 682–689.
50. Chander, Y.; Goyal, S.M.; Gupta, S.C. Antimicrobial resistance of *Providencia* spp. isolated from animal manure. *Vet. J.* 2006, *172*, 188–191.
51. Botts, C.W.; Ferguson, L.C.; Birkeland, J.M.; Winter, A.R. The influence of litter on the control of salmonella infections in chicks. *Am. J. Vet. Res.* 1952, *13*, 562–565.
52. Tucker, J.F. Survival of salmonellae in built-up litter for housing of rearing and laying fowls.
53. *Br. Vet. J.* 1967, *123*, 92–103.
54. Himathongkham, S.; Riemann, H. Destruction of *Salmonella typhimurium, Escherichia coli* O157:H7 and *Listeria monocytogenes* in chicken manure by drying and/or gassing with ammonia. *FEMS Microbiol. Lett.* 1999, *171*, 179–

182.

55. Fenlon D.R.; Ogden I.D.; Vinten, A.; Svoboda, I. The fate of Escherichia coli and *E. coli* O157 in cattle slurry after application to land. *J. Appl. Microbiol.* 2000, *88*, 149S–156S.

56. U. S. Food and Drug Administration (USFDA). Standards for the Growing, Harvesting, Packing, and Holding of Produce for Human Consumption (Proposed Rule). *FDA Food Safety Modernization Act (FSMA)*; U. S. Food and Drug Administration (USFDA): Silver Spring, MD, USA, 2013.

57. Islam, M.; Morgan, J.; Doyle, M.P.; Phatak, S.C.; Millner, P.; Jiang, X.P. Fate of *Salmonella enterica* serovar Typhimurium on carrots and radishes grown in fields treated with contaminated manure composts or irrigation water. *Appl. Environ. Microbiol.* 2004, *70*, 2497–2502.

58. Centers for Disease Control and Prevention (CDC). Surveillance for foodborne disease outbreaks-United States, 2009–2010. *Morbid. Mortal. Weekly Rep. (MMWR)* 2013, *62*, 41–47.

59. Pugh, D.G.; Rankins, D.L.; Powe, T.; D'Andrea, G. Feeding broiler litter to beef cattle. *Vet. Med.* 1994, *89*, 661–664.

60. Pugh, D.G.; Wenzel, J.G.W.; D'Andrea, G. A survey on the incidence of disease in cattle fed broiler litter. *Vet. Med.* 1994, *89*, 665–667.

61. Man Claims Poultry Litter Caused Illness, Food Safety News. Available online:

62. http://www.foodsafetynews.com/2009/12/another-poultry-litter-case-to-federal-court/#.UfCZuxY VuqB (accessed on 8 January 2014).

63. U. S. Food and Drug Administration (USFDA). *FDA Approved Animal Drug Products (Green Book)*; U. S. Food and Drug Administration (USFDA): Silver Spring, MD, USA, 2004.

64. Berry, E.D.; Woodbury, B.L.; Nienaber, J.A.; Eigenberg, R.A.; Thurston, J.A.; Wells, J.E. Incidence and persistence of zoonotic bacterial and protozoan pathogens in a beef cattle feedlot runoff control-vegetative treatment system. *J. Environ. Qual.* 2007, *36*, 1873–1882.

65. Thurston-Enriquez, J.A.; Gilley, J.E.; Eghball, B. Microbial quality of runoff following land application of cattle manure and swine slurry. *J. Water Health* 2005, *3*, 157–171.

66. Sistani, K.R.; Bolster, C.H.; Way, T.R.; Tobert, H.A.; Pote, D.H.; Watts, D.B. Influence of poultry litter application methods on the longevity of nutrient and *E. coli* in runoff from tall fescue pasture. *Water Air Soil Pollut.* 2010, *206*, 3–12.

67. Singh, R.; Kim, J.; Marion, W.S., Jr.; Luo, F.; Jiang, X. Determining thermal inactivation of *Escherichia coli* O157:H7 in fresh compost by simulating early phases of the composting process. *Appl. Environ. Microbiol.* 2011, *77*, 4126–4135.

68. Bernal, M.P.; Alburquerque, J.A.; Moral, R. Composting of animal manures and chemical criteria for compost maturity assessment. A review. *Bioresource Technol.* 2009, *100*, 5444–5453.

69. Millner, P.D. Manure Management. In *The Production Contamination Problem*; Matthews, K., Solomon, E., Sapers, G., Eds.; U.S. Department of Agriculture (USDA): Beltsville, MD, USA, 2009; Volume 1, pp. 79–104.

70. Barker, K.J.; Purswell, J.L.; Davis, J.D.; Parker, H.M.; Kidd, M.T.; McDaniel, C.D.; Kiess, A.S. Distribution of bacteria at different poultry litter depths. *Int. J. Poult. Sci.* 2010, *9*, 10–13.

71. U.S. Department of Agriculture (USDA). *Chapter 2, Composting. Part 637 Environmental Engineering, National Engineering Handbook*; U.S. Department of Agriculture (USDA): Washington, DC, USA, 2000.

72. Sweeten, J.M. Composting Manure and Sludge. In Proceedings of the National Poultry Waste Management Symposium, Columbus, OH, USA, 18–19 April 1988; Ohio State University: Columbus, OH, USA, 1988.

73. Moore, P.A.; Daniel, T.C.; Edwards, D.R. Reducing phosphorus runoff and inhibiting ammonia loss from poultry manure with aluminum sulfate. *J. Environ. Qual.* 2000, *29*, 37–49.

74. Moore, P.A.; Daniel, T.C.; Gilmour, J.T.; Shreve, B.R.; Edwards, D.R.; Wood, B.H. Decreasing metal runoff from poultry litter with aluminium sulfate. *J. Environ. Qual.* 1998, *27*, 92–99.

75. Ryckeboer, J.; Mergaert, J.; Coosemans, J.; Deprins, K.; Swings, J. Microbiological aspects of biowaste during composting in a monitored compost bin. *J. Appl. Microbiol.* 2003, *94*, 127–137.

76. Hassen, A.; Belguith, K.; Jedidi, N.; Cherif, A.; Cherif, M.; Boudabous, A. Microbial characterization during composting of municipal solid waste. *Bioresour. Technol.* 2001, *80*, 217–225.

77. Erickson, M.C.; Liao, J.; Ma, L.; Jiang, X.; Doyle, M.P. Inactivation of *Salmonella* spp. in cow manure composts formulated to different initial C:N ratios. *Bioresour. Technol.* 2009, *100*, 5898–5903.

78. Talaro, K.P.; Talaro, A. Physical and Chemical Control of Microbes. In *Foundations in Microbiology*, 4th ed.; The McGraw-Hill Companies: New York, NY, USA, 2002; pp. 325–326.

79. U. S. Food and Drug Administration (USFDA). *National Organic Standards Board Crops Committee Rcommendation for Guidance Use of Compost, Vermicompost, Processed Manure, and Compost Teas*; U. S. Food and Drug Administration (USFDA): Silver Spring, MD, USA, 2006.

80. U. S. Environmental Protection Agency (USEPA). *Control of Pathogens and Vector Attraction in Sewage Sludge*; U. S. Environmental Protection Agency (USEPA): Cincinnati, OH, USA, 2003.

81. McCaskey, T.A. *Dead Bird Composting*; Auburn University: Auburn, AL, USA, 1993.

82. Sims, J.T.; Murphy, D.W.; Handwerker, T.S. Composting of poultry wastes: Implications for dead poultry disposal and manure management. *J. Sustain. Agric.* 1993, *2*, 67–82.

83. Tiquia, S.M.; Tam, N.F.Y. Characterization and composting of poultry litter in forced-aeration piles. *Process Biochem.* 2002, *37*, 869–880.

84. Brodie, H.L.; Donald, J.O.; Conner, D.E.; Tucker, J.K.; Harkin, H.D. Field evaluation of mini-composting of poultry carcasses. *Poult. Sci.* 1994, *73*, 41.

85. Macklin, K.S.; Hess, J.B.; Bilgili, S.F. In-house windrow composting and its effects on foodborne pathogens. *J. Appl. Poult. Res.* 2008, *17*, 121–127.

86. Silva, M.E.; Lemos, L.T.; Cunha-Queda, A.C.; Nunes, O.C. Co-composting of poultry manure with low quantities of carbon-rich materials. *Waste Manag. Res.* 2009, *27*, 119–128.

87. Guan, J.; Spencer, J.L.; Sampath, M.; Devenish, J. The fate of a genetically modified *Pseudomonas* strain and its transgene during the composting of poultry manure. *Can. J. Microbiol.* 2004, *50*, 415–421.

88. Hutchison, M.L.; Walters, L.D.; Avery, S.M.; Moore, A. Decline of zoonotic agents in livestock waste and bedding heaps. *J. Appl. Microbiol.* 2005, *99*, 354–362.

89. Erickson, M.C.; Liao, J.; Boyhan, G.; Smith, C.; Ma, L.; Jiang, X.; Doyle, M.P. Fate of manure-borne pathogen surrogates in static composting piles of chicken litter and peanut hulls. *Bioresour. Technol.* 2010, *101*, 1014–1020.

90. Wichuk, K.M.; McCartney, D. A review of the effectiveness of current time-temperature regulations on pathogen inactivation during composting. *J. Environ. Eng. Sci.* 2007, *6*, 573–586.

91. Wesche, A.M.; Gurtler, J.B.; Marks, B.P.; Ryser, E.T. Stress, sublethal injury, resuscitation, and virulence of bacterial foodborne pathogens. *J. Food Protect.* 2009, *72*, 1121–1138.

92. Shepherd, M.W.; Singh, R.; Kim, J.; Jiang, X. Effect of heat-shock treatment on the survival of *Escherichia coli* O157: H7 and *Salmonella enteric* Typhimurium in dairy manure co-composted with vegetable wastes under field conditions. *Bioresour. Technol.* 2010, *101*, 5407–5413.

93. Farber, J.M.; Brown, B.E. Effect of prior heat shock on heat resistance of *Listeria monocytogenes* in meat. *Appl. Environ. Microbiol.* 1990, *56*, 1584–1587.

94. Berk, P.A.; de Jonge, R.; Zwietering, M.H.; Abee, T.; Kieboom, J. Acid resistance variability among isolates of *Salmonella enterica* serovar Typhimurium DT 104. *J. Appl. Microbiol.* 2005, *99*, 859–866.

95. Johnson, K.M.; Busta, F.F. Detection and Enumeration of Injured Bacterial Spores in Processed Foods. In *The Revival of Injured Microbes*; Andrew, M.H.E., Russell, A.D., Eds.; Academic Press: London, UK, 1984; pp. 241–256.

96. Rowe, M.T.; Kirk, R.B. Effect of nutrient starvation on the resistance of *Escherichia coli* O157:H7 to subsequent heat stress. *J. Food Protect.* 2000, *63*, 1745–1748.

97. Humphrey, T. *Salmonella*, stress response and food safety. *Nat. Rev.* 2004, *2*, 504–509.

98. Humphrey, T.J.; Williams, A.; McAlpine, K.; Lever, M.S.; Guard-Petter, J.; Cox, J.M. Isolates of *Salmonella enterica* Enteritidis PT_4 with enhanced heat and acid tolerance are more virulent in mice and moreinvasive in chickens. *Epidemiol. Infect.* 1996, *177*, 79–88.

99. Foster, J.W.; Spector, M.P. How *Salmonella* survive against the odds. *Annu. Rev. Microbiol.* 1995, *49*, 145–174.
100. Yousef, A.E.; Courtney, P.D. Basics of Stress Adaptation and Implications in New-Generation Foods. In *Microbial Stress Adaptation and Food Safety*; Yousef, A.E., Juneja V.J., Eds.; CRC Press: Boca Raton, FL, USA, 2003; pp. 1–30.
101. Abee, T.; Wouters, J.A. Microbial stress response in minimal processing. *Int. J. Food Microbiol.* 1999, *50*, 65–91.
102. Suh, S.; Silo-Suh, L.; Woods, D.E.; Hassett, D.J.; West, S.E.H.; Ohman, D.E. Effect of *rpoS* mutation on the stress response and expression of virulence factors in *Pseudomonas aeruginosa*. *J. Bacteriol.* 1999, *181*, 3890–3897.
103. Cheville, A.M.; Arnold, K.W.; Buchrieser, C.; Cheng, C.M.; Kaspar, C.W. *rpoS* Regulation of acid, heat, and salt tolerance in *Escherichia coli* O157:H7. *Appl. Environ. Microbiol.* 1996, *62*, 1822–1824.
104. Van Hoek, A.H.A.M.; Aarts, H.J.M.; Bouw, E.; van Overbeek W.M.; Franz, E. The role of *rpoS* in *Escherichia coli* O157 manure-amended soil survival and distribution of allelic variations among bovine, food and clinical isolates. *FEMS Microbiol. Lett.* 2013, *338*, 18–23.
105. Ghaly, A.E.; Alhattab, M. Drying poultry manure for pollution potential reduction and production of organic fertilizer. *Am. J. Environ. Sci.* 2013, *9*, 88–102.
106. Messer, J.W.; Lovett, J.; Murthy, G.K.; Murthy, G.K.; Wehby, A.J.; Schafer, M.L.; Read, R.B.
107. An assessment of some public health problems resulting from feeding poultry litter to animals. *Poult. Sci.* 1971, *50*, 874–881.
108. Chen, Z.; Diao, J.; Dharmasena, M.; Ionita, C.; Jiang, X.; Rieck, J. Thermal inactivation of desiccation-adapted *Salmonella* spp. in aged chicken litter. *Appl. Environ. Microbiol.* 2013, *79*, 7013–7020.
109. National Organic Standards Board. *Crops Committee Recommendation for Guidance Use of Compost, Vermicompost, Processed Manure, and Compost Teas*; National Organic Standards Board: Washington, DC, USA, 2006.
110. Commission Regulation (EC) No 208/2006 of 7 February 2006: Amending Annexes VI and VIII to Regulation (EC) No 1774/2002 of the European Parliament and of the Council as Regards Processing Standards for Biogas and Composting Plants and Requirements for Manure. Available online: http://eur-lex.europa.eu/LexUriServ/LexUriServ.do?uri=OJ:L:2006:036:0025:0031:EN:PDF (accessed on 8 January 2014).
111. California Leafy Green Products Handler Marketing Agreement. *Commodity Specific Food Safety Guidelines for the Production and Harvest of Lettuce and Leafy Greens*; California Leafy Green Products Handler Marketing Agreement: Sacramento, CA, USA, 2010.
112. Barbour, E.K.; Husseini, S.A.; Farran, M.T.; Itani, D.A.; Houalla, R.H.; Hamadeh, S.K. Soil solarization: A sustainable agriculture approach to reduce microorganisms in chicken manure-treated soil. *J. Sustain. Agric.* 2002, *19*, 95–104.

113. Oni, R.; Sharma, M.; Micallef, S.; Buchanan, R. The effect of UV radiation on survival of *Salmonella enterica* in dried manure dust. In Proceedings of the International Association for Food Protection Annual Meeting, Charlotte, NC, USA, 30 July 2013.

114. Rothrock, M.J.; Cook, K.L.; Warren, J.G.; Sistani, K. The effect of alum addition on microbial communities in poultry Litter. *Poult. Sci.* 2008, *87*, 1493–1503.

115. Gandhapudi, S.K.; Coyne, M.S.; D'Angelo, E.M.; Matocha, C. Potential nitrification in alum-treated soil slurries amended with poultry manure. *Bioresour. Technol.* 2006, *97*, 664–670.

116. Scantling, M.; Waldroup, A.; Mary, J.; Moore, P. Microbiological effects of treating poultry litter with aluminum sulfate. *Poult. Sci. (Abstr.)* 1995, *74*, 216.

117. Line, J.E. Aluminum sulfate treatment of poultry litter to reduce *Salmonella* and *Campylobacter* populations. *Poult. Sci. (Abstr.)* 1998, *77*, S364.

118. Line, J.E. *Campylobacter* and *Salmonella* populations associated with chicken raised on acidified litter. *Poult. Sci.* 2002, *81*, 1473–1477.

119. U. S. Environmental Protection Agency (USEPA). *Biosolids Generation, Use, and Disposal in the United States*; U. S. Environmental Protection Agency (USEPA): Cincinnati, OH, USA, 1999.

120. Stringfellow, K.; Caldwell, D.; Lee, J.; Byrd, A.; Carey, J.; Kessler, K.; Farnell, M.

121. Pasteurization of chicken litter with steam and quicklime to reduce *Salmonella* Typhimurium. *J. Appl. Poult. Res.* 2010, *19*, 380–386.

122. Hills, B.P.; Manning, C.E.; Ridge, Y.; Brocklehurst, T. Water availability and the survival of *Salmonella* typhimurium in porous systems. *Int. J. Food Microbiol.* 1997, *36*, 187–198.

123. Maguire, R.O.; Hesterberg, D.; Gernat, A.; Anderson, K.; Wineland, M.; Grimes, J. Liming poultry manures to decrease soluble phosphorus and suppress the bacteria population. *J. Environ. Qual.* 2006, *35*, 849–857.

124. Bennett, D.S.; Higgins, S.E.; Moore, R.; Beltran, R.; Caldwell, D.; Byrd, J.A.; Hargis, B.M. Effects of lime on *Salmonella enteritidis* survival *in vitro*. *J. Appl. Poult. Res.* 2003, *12*, 65–68.

125. Bennett, D.S.; Higgins, S.E.; Moore, R.; Byrd, J.A.; Beltran, R.; Corsiglia, C.; Caldwell, D.; Hargis, B.M. Effect of addition of hydrated lime to litter on recovery of selected bacteria and poult performance. *J. Appl. Poult. Res.* 2005, *14*, 721–727.

126. Turnbull, P.C.; Snoeyenbos, G.H. The roles of ammonia, water activity, and pH in the salmonellacidal effect of long-used poultry litter. *Avian Dis.* 1973, *17*, 72–86.

127. Wang, G.; Zhao, T.; Doyle, M.P. Fate of enterohemorrhagic *Escherichia coli* O157:H7 in bovine feces. *Appl. Environ. Microbiol.* 1996, *62*, 2567–2570.

128. Ferguson, N.S.; Gates, R.S.; Taraba, J.L.; Cantor, A.H.; Pescatore, A.J.; Straw, M.L.; Ford, M.J.; Burnham, D.J. The effect of dietary protein and phosphorus on ammonia concentration and litter composition in broilers. *Poult. Sci.* 1998,

77, 1085–1093.

129. Fontenot, J.P.; Webb, K.E.; Harmon, B.W.; Tucker R.E.; Moore, W.E.C. Studies of processing nutritional value and palatability of broiler litter for ruminants. In Proceedings of the International Symposium on Livestock Wastes, Columbus, OH, USA, 19–22 April 1971.
130. Caswell, L.F.; Fontenot, J.P.; Webb, K.E. Effect of processing treatment on pasteurization and nitrogen components of broiler litter and on nitrogen utilization by sheep. *J. Anim. Sci.* 1975, *4*, 750–758.
131. Seltzer, W.; Moum, S.G.; Goldhart, T.M. A method for the treatment of animal wastes to control ammonia and other odors. *Poult. Sci.* 1969, *48*, 1912–1918.
132. Murray, G.E.; Tobin, R.S.; Junkins, B.; Kushner, D.J. Effect of chlorination on antibiotic resistance profiles of sewage-related bacteria. *Appl. Environ. Microbiol.* 1984, *48*, 173–177.
133. Munir, M.; Wong, K.; Xagoraraki, I. Release of antibiotic resistant bacteria and genes in the effluent and biosolids of five wastewater utilities in Michigan. *Water Res.* 2011, *45*, 681–693.
134. Chen, Z.; Zhu, C.; Han, Z. Effects of aqueous chlorine dioxide treatment on nutritional components and shelf-life of mulberry fruit (*Morus alba* L.). *J. Biosci. Bioeng.* 2011, *111*, 675–681.
135. Erickson, M.C.; Islam, M.; Sheppard, C.; Liao, J.; Doyle, M.P. Reduction of *Escherichia coli* O157:H7 and *Salmonella enterica* serovar Enteritidis in chicken manure by larvae of the black soldier fly. *J. Food Protect.* 2004, *67*, 685–690.
136. Yongabi, K.A.; Harris, P.L.; Lewis, D.M. Poultry faeces management with a simple low cost plastic digester. *Afr. J. Biotechnol.* 2009, *8*, 1560–1566.
137. Krylova, N.I.; Khabiboulline, R.E.; Naumova, R.P.; Nagle, M. The influence of ammonium and methods for removal during the anaerobic treatment of poultry manure. *J. Chem. Technol. Biotechnol.* 1997, *70*, 99–105.
138. Heringa, S.D.; Kim, J.; Jiang, X.; Doyle, M.P.; Erickson, M.C. Use of a mixture of bacteriophages for biological control of *Salmonella enteric* strains in compost. *Appl. Environ. Microbiol.* 2010, *76*, 5327–5332.

Chapter 4

PERFORMANCE OF MAIZE LANDRACE UNDER NO-TILL AS AFFECTED BY THE ORGANIC AND MINERAL FERTILIZERS

Eduardo Fávero Caires*; Helio Antonio Wood Joris; Susana Churka; Renato Zardo Filho

Universidade Estadual de Ponta Grossa; Av. Gal. Carlos Cavalcanti 4748; 84030-900; Ponta Grossa - PR - Brasil

ABSTRACT

The aim of this work was to study the effects of organic and mineral fertilizers at sowing (without fertilizers, organic poultry litter fertilizer on the surface and mineral NK + reactive natural phosphate from Arad and NK + triple superphosphate in the furrow) and topdressing (without fertilizers, organic poultry litter fertilization and urea) on chemical attributes of a no-till Oxisol and nutrition and yield of maize landrace (Zea mays L.), Carioca variety in a field experiment. Results revealed that P content (Mehlich 1 and resin) was increased in the soil surface layer with organic poultry litter fertilizer on the surface at sowing. Mineral fertilizer in the sowing furrow could be replaced by organic fertilizer with poultry litter on the surface, but topdressing fertilization with urea resulted better N nutrition for the plants and higher grains yield than the organic poultry litter fertilization.

INTRODUCTION

No-till systems are as one of the most effective strategies to improve the sustainability of farming system in the tropical and subtropical regions by reducing soil and nutrient losses through erosion (Lal 1995; Hobbs et al. 2008). The growth of the area under no-till has been especially rapid in South America where some countries such as Argentina, Brazil, Paraguay, and Uruguay are using this system on about 70% of the total cultivated agricultural area (Derpsch and Friedrich 2009). In Brazil, the cultivated agricultural area under no-till has rapidly increased to 25.5 million hectares (Derpsch and Friedrich 2009), including the areas in which conventional crop management systems and natural pasture grazing by the cattle are common.

With the evolution of agriculture, many advances related to the crop management and the development of genotypes with high productive potential adapted to different environments have been achieved. Because the search for higher grain yield originated great changes in relation to the genetic interaction with management techniques, over the years, maize crop is often regarded as a kind of technological development model.

Landraces have been gradually substituted by maize hybrids with higher productive potential, which resulted in genetic erosion of maize landraces (Paterniani 1980). Despite being less productive than the maize hybrid varieties, the landraces present high potential of adaptation to soil-climate conditions, lower cost to obtain the seeds and high crop potential to attend market niches (Paterniani et al. 2000; Pinto et al. 2009). However, very little information is known about their nutritional demands for agronomic adaptation to different crop environments.

Fertilizers cost in Brazil has contributed significantly to the reduction of crop profits. Organic fertilizers might become a viable alternative for the maize landrace production by small farmers, considering the high cost of mineral fertilizers and the possibility of aggregating value to the product through quality improvement.

Poultry litter is one of the best organic fertilizers available for agriculture uses as it shows the ability to increase the organic matter and supply, mainly, nitrogen (N) and phosphorus (P) to the plants (Moore et al. 1995; Endale et al. 2008; Yadvinder-Singh et al. 2009).

Reactive rock phosphates (sedimentary rocks) might also be an interesting option for small farmers in replacement to the totally acidulated phosphates, as they usually represent lower cost per P unit. They are usually less soluble in water and effective within specific conditions of soil management, such as high organic matter and clay content acid soils, with high P adsorption capability (Chien and Menon 1995; Prochnow et al. 2006; Pauletti et al. 2010).

Despite increasing interest in substituting the mineral fertilizers by organic fertilizers and fast growth of no-till crop areas in Brazil, effects of organic fertilizers on the surface, both in sowing and topdressing, aiming at replacing or complementing mineral fertilization in maize crops are unknown. This work examined the effects of organic and mineral fertilizers at sowing and topdressing on soil chemical attributes and grain yield of maize landrace crop under a no-till system.

MATERIALS AND METHODS

The experiment was carried out at Ponta Grossa, Paraná State, Brazil (25°10'S, 50°05'W), on a loamy Typic Hapludox under a no-till system for eight years. Table 1 shows the soil chemical analyses results, layers 0-10 to 10-20 cm, carried out before the beginning of the experiment. A randomized complete block design was used, with three replications in a 4 × 3 factorial arrangement. The 38.4 m^2 (6.4 × 6.0 m) plots received four treatments of fertilizers at sowing: no fertilization, organic fertilization, NK + reactive natural phosphate from Arad (RNF) and NPK mineral fertilization, as well as three treatments of topdressing fertilization: no fertilization, organic fertilization and nitrogen mineral fertilization. Based on the results of the soil chemical analyses and the maize nutritional demands, 30 kg ha^{-1} of N, 60 kg ha^{-1} of P_2O_5 and 45 kg ha^{-1} of K_2O were applied at sowing and 120 kg ha^{-1} of N at topdressing.

Organic fertilization at sowing and topdressing was carried out by applying the poultry litter on the soil surface at 10 and 5 t ha^{-1} respectively, based on dry weight and organic fertilizer nutrients (Table 2). The rates of poultry litter used at sowing and topdressing were calculated from the maize crop need of N, considering about 50% mineralization (CQFS - RS/SC, 1997) of the total N found in the compound. Nitrogen, P, and K mineral sources used were urea, triple superphosphate (TSP) and potassium chloride (KCl), respectively. The topdressing fertilization was carried out at the vegetative stage V_4.

Table 1 - Soil chemical attributes at 0–10 and 10–20 cm depths, before the establishment of the experiment.

Depth	pH ($CaCl_2$)	Al^{3+}	H + Al	Ca^{2+}	Mg^{2+}	K^+	P (Mehlich 1)	Organic–C
cm		$mmol_{(+)}$ dm^{-3}					mg dm^{-3}	g dm^{-3}
0–10	5.1	0.0	53.5	28	17	2.9	12.8	28
10–20	5.1	0.0	53.5	24	16	1.7	2.6	18

Table 2 - Characteristics of poultry litter used in the study.

Water content	Organic–C	N	P	K	C/N ratio
g kg^{-1}					
389.4	194.1	16.3	11.4	16.7	11.9

The maize (*Zea mays* L.), variety Carioca, was sowed in November 2006, after black oat (*Avena strigosa* Schreb) growth at a seeding rate of 6 seeds/m, and row spacing of 0.80 m. The landrace Carioca presented dent grains and high potential of response to fertilization when compared to other varieties (Machado et al. 1999). Seasonal rainfall and air temperature data for the period 2006-07 when the maize crop was in the field are shown in Figure 1. Throughout the development period of the crop, weather conditions were normal and there was no water limitation.

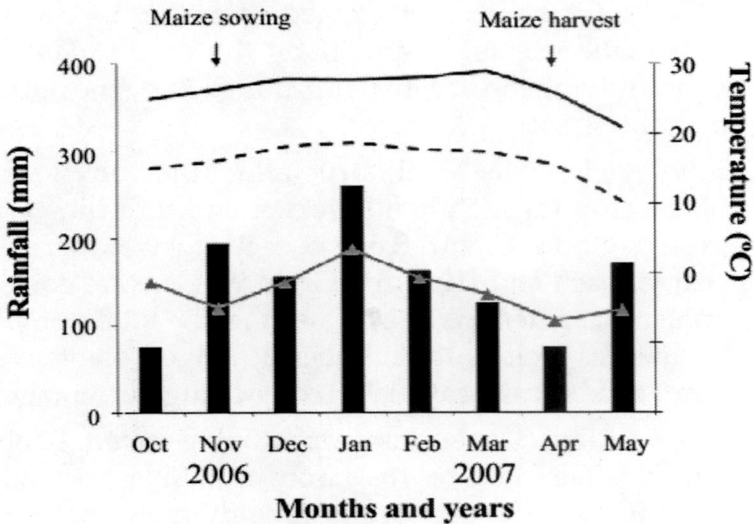

Figure 1 – Monthly rainfall (vertical bars) and minimum (---) and maximum (–) temperatures during the maize crop development period, and 45 years average monthly rainfall at Ponta Grossa, southern Brazil.

Samples of leaves were collected at the beginning of the maize crop flowering. The leaf under and opposite to the ear was picked from 30 plants in each plot. Chlorophyll content was measured in the leaves collected, with ten readings of the central part of each leaf, using a Minolta SPAD-502 chlorophyllometer. After that, the leaves samples were washed with deionized water, the leaves ends were cut out and the veins removed.

The leaves samples that contained the mid-third without veins were dried in a forced-air oven at 60 °C until constant mass was achieved and ground. Nitrogen, P, and K contents were determined according to the methods described by Malavolta et al. (1997).

Ten plants from each plot were used for measurement of culm diameter and height carried out at the flowering peak. Culm diameter was measured with a pachymeter and the plants height with a ruler, from the plant base to the last culm node. At the same time, ten plants were randomly removed from each plot. Plants collected were then washed with deionized water and dried in oven with forced air circulation at 60 °C until the constant mass was obtained. After dry matter production evaluation, the plants were ground and after that N, P and K contents were determined according to Malavolta et al. (1997). Extraction of N, P, and K by the maize plants was calculated by multiplying the dry matter production by the nutrients concentration in the shoot.

Maize grain yield was evaluated after the physiological maturation of the crop through hand-picking and threshing with a stationary thresher. The four central rows of each plot were harvested, discharging 1 m of each end (12.8 m^2). Grain yields were corrected to 130 g kg^{-1} moisture content. While harvesting, the total number of plants was counted as well as the number of lodged plants within the area (12.8 m^2) used for the calculation of lodging percentage.

Soil samples were taken as soon as maize was harvested. To obtain a composite sample, five soil cores (two from each inter row and one from the sowing row) were taken from three different locations in each plot at 0-5, 5-10 and 10-20 cm depth using a soil probe. Organic carbon content was determined through the Walkley-Black method. Exchangeable K^+ was extracted with Mehlich 1 solution (Pavan et al. 1992) and P with Mehlich 1 solution (Pavan et al. 1992) and anionic exchange resin (Raij et al. 2001).

Results were submitted to the analysis of variance, following a randomized complete block design in a factorial arrangement model. Treatment effects were compared by the Tukey test at $p = 0.05$.

RESULTS AND DISCUSSION

Variance analysis of the chemical analyses of soil did not reveal a significant interaction between the fertilization treatments at sowing and topdressing. The organic C and exchangeable K^+ contents, at different soil depths were not influenced by the organic or mineral fertilization at sowing or topdressing (Table 3). Soil P content extracted with Mehlich 1 solution or resin was significantly higher at the soil surface layer (0-5cm) when organic fertilization at sowing was applied, when compared with the treatments without fertilization or with NK + RNF and NK + TSP fertilization.

Table 3 – Contents of organic C, exchangeable K, and available P in the soil as affected by the fertilizer treatments at sowing and topdressing for growing the maize landrace under a no-till system

Treatment	Organic-C	K^+	P	
			Mehlich 1	Resin
	g dm^{-3}	mmol$_{(+)}$ dm^{-3}	--------- mg dm^{-3} ---------	
		0–5 cm		
Sowing fertilizer				
No fertilizer	22.9 a	4.8 a	18.1 b	22.6 b
Poultry litter	23.2 a	5.6 a	38.5 a	34.8 a
NK + Arad Phosphate Rock	22.2 a	4.7 a	18.5 b	21.1 b
NK + Triple Superphosphate	22.3 a	4.8 a	17.4 b	20.5 b
Topdressing fertilizer				
No fertilizer	21.9 a	5.1 a	23.2 a	21.1 b
Poultry litter	22.8 a	4.9 a	22.4 a	28.1 a
Urea	23.2 a	5.0 a	23.9 a	25.0 ab
CV (%)	7.2	15.4	31.1	21.0
		0–10 cm		
Sowing fertilizer				
No fertilizer	19.5 a	3.7 a	17.5 a	17.9 bc
Poultry litter	20.5 a	4.2 a	22.8 a	24.3 a
NK + Arad Phosphate Rock	20.4 a	3.6 a	15.5 a	15.4 c
NK + Triple Superphosphate	20.6 a	3.9 a	18.7 a	21.0 ab
Topdressing fertilizer				
No fertilizer	20.4 a	3.8 a	17.9 a	19.3 a
Poultry litter	20.0 a	3.9 a	18.4 a	19.6 a
Urea	20.3 a	3.8 a	19.6 a	20.1 a
CV (%)	9.4	13.1	30.4	18.5
		0–20 cm		
Sowing fertilizer				
No fertilizer	19.1 a	2.6 a	9.5 a	11.9 b
Poultry litter	18.7 a	3.1 a	10.2 a	14.9 a
NK + Arad Phosphate Rock	19.0 a	2.6 a	7.4 a	10.2 b
NK + Triple Superphosphate	18.8 a	2.7 a	6.9 a	9.6 b
Topdressing fertilizer				
No fertilizer	18.9 a	2.9 a	8.1 a	9.9 a
Poultry litter	18.5 a	2.8 a	9.1 a	11.5 a
Urea	19.2 a	2.6 a	8.4 a	13.6 a
CV (%)	7.5	16.3	34.8	25.0

Anionic exchange resin presented higher sensitiveness than Mehlich 1 solution to detect the increase in soil P at 0-10 and 0-20 cm depths, with organic fertilization at sowing, and also at 0-5 cm depth, with topdressing organic fertilization. Soil P-resin content at 0-10 cm depth was similar with the use of organic fertilization or NK + TSP at sowing. However, the higher increase in soil P was obtained with the addition of poultry litter on the soil surface at sowing. According to the available P found in the soil, compared to the control treatment, it could be observed that there was high P mineralization added by the poultry litter. Besides, due to the presence of calcium and magnesium phosphates and the formation of precipitates in the poultry litter (Fordham and Schwertmann 1977), continuous P release occurs during the growing time because such compounds tend to react more slowly than the soluble fertilizers (Sharpley and Sisak 1997).

This effect was also expected with the application of RNF; however, the P rate added with the use of RNF was smaller than the one applied through poultry litter. Considering the TSP application, its high solubility might have resulted in an increase in P availability at the early development of the maize crop; however, at the time of the soil sampling, there was no difference in relation to the treatment without fertilization. It should be considered that the soil had sufficient content of P at the 0-10 cm depth (Table 1).

Leaf content and uptake of N, P, and K by maize landrace plants were not significantly influenced by the interaction between the fertilizers treatments at sowing and topdressing. Organic fertilizer applied during the maize sowing provided better N leaf content than the treatments without fertilizers and with NK + RNF and NK + TSP fertilization (Table 4).

Nitrogen extraction by the shoot of maize was higher after organic fertilization or NK + TSP application than after NK + RNF application or without fertilizers at sowing. Such effects should be regarded as a result of N × P interaction, since N extraction by the shoot was not significantly different in the treatments without fertilizers or with NK + RNF at sowing. Thus, higher soil P content provided by the organic fertilizaers and TSP addition might have favored N uptake by the maize plants. Alves et al. (1999) also observed higher N accumulation in the shoot of maize when P and N were supplied together to all root system.

The increase in P concentration in the maize leaf tissue resulting from the organic fertilization at sowing (Table 4) was followed by the increase in soil P availability, extracted by Mehlich 1 solution or resin (Table 3). It was observed that exchangeable K^+ content in the soil was not limiting and the treatments with organic or mineral fertilization at sowing did not exert any influence on the K^+ content in the soil (Table 3) or in the maize leaf tissue (Table 4). On the other hand, N extraction by the shoot and N, P and K concentrations in the maize leaves were significantly higher with urea application than with organic fertilizer in topdressing (Table 4). Therefore, organic fertilization with poultry litter proved efficient in providing adequate nutrition to maize, when employed to replace the mineral fertilization at sowing, but urea provided a better plant nutrition than the poultry litter fertilization in topdressing.

Table 4 – Leaf content and uptake of N, P, and K by maize landrace plants as affected by fertilizer treatments at sowing and topdressing under a no-till system

Treatment	Leaf content			Nutrient uptake		
	N	P	K	N	P	K
	---------- g kg^{-1} ----------			---------- kg ha^{-1} ----------		
Sowing fertilizer						
No fertilizer	25.1 b	2.8 b	20.9 a	242.2 b	9.5 a	146.3 a
Poultry litter	28.3 a	3.4 a	21.9 a	337.7 a	12.3 a	171.8 a
NK + Arad Phosphate Rock	26.1 ab	2.9 b	21.8 a	301.0 ab	9.7 a	173.8 a
NK + Triple Superphosphate	26.2 ab	2.9 b	20.7 a	331.8 a	10.2 a	188.3 a
Topdressing fertilizer						
No fertilizer	24.2 b	2.8 b	21.3 ab	263.8 b	11.1 a	149.7 a
Poultry litter	24.9 b	2.9 b	20.5 b	277.9 b	11.4 a	176.4 a
Urea	32.2 a	3.7 a	22.2 a	367.9 a	8.8 a	184.2 a
CV (%)	7.7	10.7	7.1	21.9	26.7	22.7

Figure 2 shows correlation data obtained between the soil P (0-20 cm) extracted by Mehlich 1 solution and resin and maize landrace leaf P contents for the treatments with organic fertilizer with poultry litter and mineral fertilizer with NK + TSP at sowing. Treatments without fertilizers and NK + RNF fertilizers at sowing did not result in soil P changes (Table 3), and, for this reason, were not taken into consideration in this analysis. The increase in soil P detected by the Mehlich 1 and resin extractants (Table 3) was correlated to P increase in the maize leaves, when organic fertilizer was surface-applied at sowing (Figure 2). However, when NK + TSP was applied in the sowing furrow, only the extraction with the resin method detected gains in soil P availability related to increase in P content in the maize leaves.

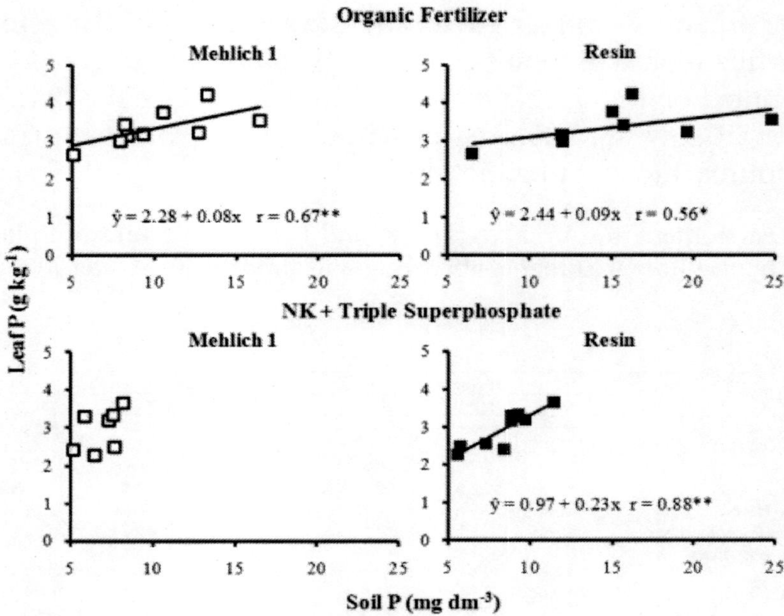

Figure 2 – Relationship between soil P (0–20 cm) extracted by Mehlich 1 solution () and resin () and maize landrace leaf P contents for the treatments with organic fertilizer with poultry litter and mineral fertilizer with NK + triple superphosphate at sowing. *: $p < 0.05$ and **: $p < 0.01$.

Mehlich 1 solution extracts P bound to Ca and, in lower proportion, P bound to Fe and Al (Raij 1991). Acid extractants have been criticized, mainly for their capability to solubilize the unavailable P to the plants in the soils that received natural phosphates application (Moreira and Malavolta 2001). Extraction of soil P by anionic exchange resin presents certain similarities with P uptake by the plants roots, characterized by labile P transference to the soil solution and from there to the roots (Silva and Raij 1999). Some studies have shown superiority of the resin method in discriminating the available P in the soils fertilized with natural phosphates (Korndörfer et al. 1999). However, soil P determination by the resin in these circumstances is also subject to problems resulting from intense soil milling, with disaggregation and particle breakage, jeopardizing the analytical results (Novais and Smyth 1999). In the present study, both the extractants, Mehlich 1 and resin, presented similar efficiency in estimating the available P after RNF application. That meant that none of the extractants detected changes in soil P content following RNF application (Table 3), but P content in the maize leaf tissue was not changed after RNF application (Table 4). A closest relationship between the soil P and leaf P contents was obtained with Mehlich 1 after organic fertilization with poultry litter and with anionic exchange resin after fertilization with TSP (Figure 2).

Chlorophyll content in the leaves (SPAD), dry biomass, height, culm diameter, and lodging of maize landrace plants were not significantly influenced by the interaction between the fertilizers treatments at sowing and topdressing. Chlorophyll reading (SPAD) in the maize leaves (Table 5) showed similar behavior as the N leaf content (Table 4) to the fertilizers treatments at sowing and topdressing. The highest chlorophyll values in the leaves were obtained with organic fertilization at sowing and urea at topdressing (Table 5). The close correlation ($r = 0.87$, $p < 0.01$) obtained between the N content (, in g kg^{-1}) and chlorophyll reading (x, in SPAD) in maize landrace leaves ($= -7.88 + 0.787x$) showed the possibility to estimate the N content in leaf tissue based on the chlorophyll meter reading. The chlorophyll relative content was based on the positive correlation between the chlorophyll and leaf N levels (Argenta et al. 2001; Rambo et al. 2004), so that the indirect measurement of chlorophyll in the leaves through portable devices in the field might provide valuable information for the management of nitrogen fertilization. NK + TSP application at sowing resulted in higher maize plants and larger dry biomass production than the treatment without fertilizers, even if the plants height and dry biomass production were similar for all the organic or mineral fertilizers treatments at sowing (Table 5). The plants culm diameter was larger in the treatments with organic or mineral fertilizers at sowing than in the treatment without the fertilizers. Regarding topdressing fertilization, the use of urea resulted in plants with increased height and larger dry biomass production than the treatment without fertilizers and plants with larger culm diameter than the treatment with organic fertilizer. The dry biomass production in the treatment with poultry litter application at topdressing was similar to the treatment without topdressing. Improvement in the plant nutrition caused by the fertilizers at sowing resulted in more vigorous plants resistant to lodging, taking into consideration that the highest percentage of lodged plants was obtained in the treatment without fertilization at sowing. The elevated percentage of lodged maize plants in the present study was a result of the high height of the landrace variety associated with the occurrence of winds in the region.

Table 5 – Chlorophyll content in leaves (SPAD), dry biomass, height, culm diameter and lodging of maize landrace plants as affected by fertilizer treatments at sowing and topdressing under a no-till system

Treatment	Chlorophyll (SPAD)	Dry biomass	Plant height	Culm diameter	Lodged plants
		kg ha^{-1}	---------- m	----------	%
Sowing fertilizer					
No fertilizer	42.0 b	9567.3 b	2.9 b	0.021 b	69.7 a
Poultry litter	45.3 a	12044.7 ab	3.0 ab	0.023 a	60.0 ab
NK + Arad Phosphate Rock	44.0 ab	11150.6 ab	3.0 ab	0.023 a	52.6 b
NK + Triple Superphosphate	44.0 ab	12462.4 a	3.1 a	0.023 a	59.5 ab
Topdressing fertilizer					
No fertilizer	41.5 b	10888.3 a	2.9 b	0.022 ab	56.1 a
Poultry litter	42.2 b	10848.5 a	3.0 ab	0.021 b	61.5 a
Urea	47.8 a	12281.9 b	3.1 a	0.023 a	63.8 a
CV (%)	3.9	12.7	5.0	7.6	16.0

Maize grain yields were significantly influenced by the interaction between the fertilizers treatments at sowing and topdressing (Fig. 3 and 4). When there was no topdressing fertilization, organic fertilization with poultry litter resulted in a higher maize yield than the treatment without fertilization at sowing (Fig. 3). Maize yield was higher with poultry litter or NK + RNF and NK + TSP in relation to the treatment without fertilization at sowing, when poultry litter was applied at topdressing (Fig. 3). Fertilization with NK + TSP resulted in a higher maize yield than the treatment without fertilization at sowing, when urea was used at topdressing (Fig. 3). Topdressing fertilization, both poultry litter and with urea, did not change maize yield in the absence of fertilization at sowing (Fig. 4). Urea application provided a higher maize yield than topdressing poultry litter fertilization, when fertilization at sowing was poultry litter or with NK + RNF and NK + TSP (Fig. 4). Results revealed that in order to obtain higher maize landrace yield, fertilization at sowing could either be organic with poultry litter or mineral with NK + RNF or NK + TSP, but topdressing fertilization should be with urea. Poultry litter application at topdressing showed low efficiency for the maize production, with similar effect to treatment without topdressing. Because the mineralization of organic N of the poultry litter occurs over time after application (Yadvinder-Singh et al. 2009), the utilization of N from poultry litter by maize crop is increased with the previous application at sowing.

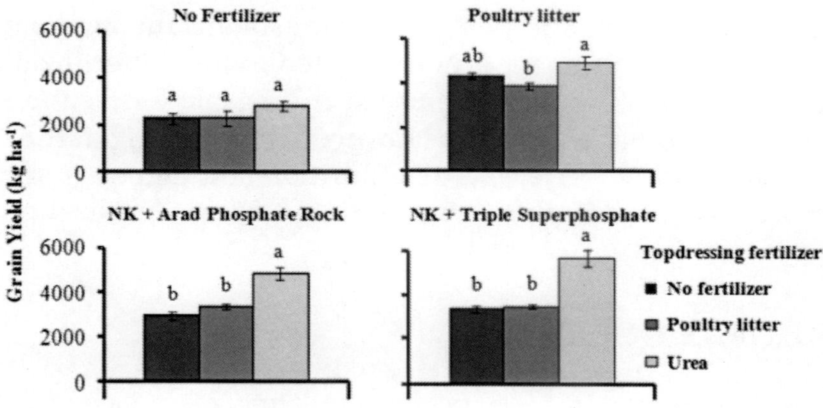

Figure 3 - Grain yield of maize landrace as affected by treatments without fertilizer and with poultry litter, NK + Arad phosphate rock, and NK + triple superphosphate at sowing, considering the treatments without fertilizer and with poultry litter, and urea at topdressing. Same letters do not differ significantly by Tukey test at p = 0.05. Vertical bars indicate standard deviation of the mean.

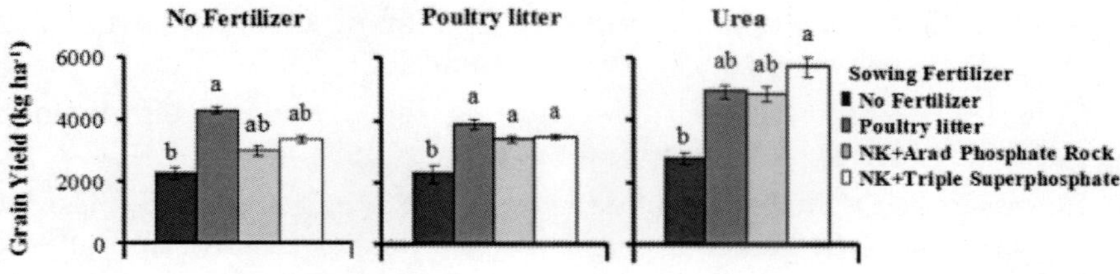

Figure 4 - Grain yield of maize landrace as affected by treatments without fertilizer and with poultry litter, and urea at topdressing, considering the treatments without fertilizer and with poultry litter, NK + Arad phosphate rock, and NK + triple superphosphate at sowing. Same letters do not differ significantly by Tukey test at p = 0.05. Vertical bars indicate standard deviation of the mean.

CONCLUSIONS

Organic fertilization with poultry litter on the surface at sowing of maize landrace under a no-till system resulted in higher available P in the soil surface layer than NK + reactive natural phosphate from Arad and NK + triple superphosphate in the sowing furrow. For maize landrace production under a no-till system, mineral fertilization in the sowing furrow could be replaced by organic fertilization with poultry litter on the surface. However, topdressing fertilization with urea provided better plant N nutrition and higher grain yield, therefore it is not advisable to replace it by organic fertilization with poultry litter.

ACKNOWLEDGEMENTS

Authors thank to the FINEP and FUNDAÇÃO ARAUCÁRIA for financial support to the research project BIOAGROPAR: Bioprospecting and technology for development of the maize landrace production chain.

REFERENCES

1. Alves VMC, Magalhães JV, Vasconcelos CA, Novais RF, Bahia Filho AFC, França GE, et al. Acúmulo de nitrogênio e de fósforo em plantas de milho afetadas pelo suprimento parcial de fósforo às raízes. *R Bras Ci Solo*. 1999; 23: 299-305
2. Argenta G, Silva PRFda, Bortolini CG. Teor de clorofila na folha como indicador do nível de N em cereais. *Ci Rural*.2001; 31: 715-722
3. Chien SH, Menon RG. Factors affecting the agronomic effectiveness of phosphate rock for direct application. *Fert Res*. 1995; 41: 227-234
4. Comissão de Química e Fertilidade do Solo RS/SC - CQFS-RS/SC. Manual de adubação e de calagem para o Estado do Rio Grande do Sul e Santa Catarina. Porto Alegre: SBCS/Núcleo Regional Sul; 2004
5. Derpsch R, Friedrich T. Global overview of conservation agriculture adoption. In: IV World Congress on Conservation Agriculture: Proceedings: Lead Papers; 2009 Feb; New Delhi, India. New Delhi: Indian Council of Agricultural Research: ICAR; 2009. p. 429-438
6. Endale DM, Schomberg HH, Fisher DS, Jenkins MB, Sharpe RR, Cabrera ML. No-Till corn productivity in a southeastern United States Ultisol amended with poultry litter. *Agron J*. 2008; 100: 1401-1408

7. Fordhan AW, Schwertmann U. Composition and reaction of liquid manure (gülle), with particular reference to phosphate - II: Solid phase components. *J Environ Qual.* 1977; 6: 136-140

8. Hobbs PR, Sayre K, Gupta R. The role of conservation agriculture in sustainable agriculture. *Philos T Roy Soc B.*2008; 363: 543-555

9. Kiehl EJ. Fertilizantes orgânicos. Piracicaba: Ceres; 1985

10. Korndörfer GH, Lara-Cabezas WA, Horowitz N. Eficiência agronômica de fosfatos naturais reativos na cultura do milho. *Sci Agric.* 2009; 56: 399-404

11. Lal R. Sustainable management of soil resources in the humid tropics. Tokyo: The United Nations University; 1995

12. Machado CTT, Guerra JGM, Almeida DL, Machado A T. Variabilidade entre genótipos de milho para eficiência no uso de fósforo. *Bragantia.* 1999; 58: 109-124

13. Malavolta E, Vitti GC, Oliveira SA. Avaliação do estado nutricional das plantas: princípios e aplicações. 2ª ed. Piracicaba: Potafós; 1997

14. Moore PA, Daniel TC, Edwards DR. Effect of chemical amendments on ammonia volatilization from poultry litter. *J Environ Qual.* 1995; 24: 293-300]

15. Moreira A, Malavolta E. Fontes, doses e extratores de fósforo em alfafa e centrosema. *Pesq Agropec Bras.* 2001; 36: 1519-1527

16. Novais RF, Smyth TJ. Fósforo em solo e planta em condições tropicais. Viçosa: Universidade Federal de Viçosa; 1999

17. Paterniani E. Melhoramento e produção de milho no Brasil. Piracicaba: Fundação Cargill; 1980

18. Paterniani E, Nass LL, Santos MX. O valor dos recursos genéticos de milho para o Brasil: uma abordagem histórica da utilização do germoplasma. In: Udry CV, Duarte W, editors. Uma história brasileira do milho: o valor dos recursos genéticos. Brasília: Paralelo 15; 2000. p. 11-41

19. Pauletti V, Serrat BM, Motta ACV, Favaretto N, Anjos A. Yield response to fertilization strategies in no-tillage soybean, corn and common bean crops. *Braz Arch Biol Technol.* 2010; 53: 563-574

20. Pavan MA, Bloch MF, Zempulsky HC, Miyazawa M, Zocoler DC. Manual de análise química do solo e controle de qualidade. Londrina: Instituto Agronômico do Paraná; 1992

21. Pinto ATB, Pereira J, Oliveira TR, Prestes RA, Mattielo RR, Demiate IM. Characterization of corn landraces planted grown in the Campos Gerais region (Parana, Brazil) for industrial utilization. *Braz Arch Biol Technol.* 2009; 52: 17-28

22. Prochnow LI, Quispe JFS, Francisco EAB, Braga G. Effectiveness of phosphate fertilizers of different water solubilities in relation to soil phosphorus adsorption. *Sci Agric.* 2006; 63: 333-340

23. Rambo L, Silva PRF, Argenta G, Sangoi L. Parâmetros de planta para aprimorar o manejo da adubação nitrogenada de cobertura em milho. *Ci Rural.* 2004; 34: 1637-1645. [Links]

24. Sharpley AN, Sisak I. Differential availibility of manure and inorganic sources of phosphorus in soil. *Soil Sci Soc Am J*. 1997; 61: 1503-1508

25. Silva FC, Raij B van. Disponibilidade de fósforo em solos avaliada por diferentes extratores. *Pesq Agropec Bras*.1999; 34: 267-288

26. Raij B van. Fertilidade do solo e adubação. São Paulo: Ceres; 1991

27. Raij B van, Andrade JC, Cantarella H, Quaggio JA. Análise química para avaliação da fertilidade de solos tropicais. Campinas: Instituto Agronômico; 2001

28. Yadvinder-Singh, Gupta RK, Thind HS, Bijay-Singh, Varinderpal-Singh, Gurpreet-Singh, et al. Poultry litter as a nitrogen and phosphorous source for the rice-wheat cropping system. *Biol Fert Soils*. 2009; 45: 701-710.

Chapter 5

NATURAL RADIOACTIVITY AND DOSE ASSESSMENT FOR BRANDS OF CHEMICAL AND ORGANIC FERTILIZERS USED IN SAUDI ARABIA

W. R. Alharbi

Physics Department, Faculty Science for Girls, king Abdulaziz University, Jeddah, KSA Email: Walhrbi@kau.edu.sa

ABSTRACT

The activity concentration of naturally occurring radionuclides 40K, 226Ra and 232Th have been measured in different brands of fertilizer samples in Saudi Arabia using sodium iodide gamma spectrometry. The results of measurements showed that the mean (ranges) of specific activities for 226Ra, 232Th and 40K activities in the nitrogen, phosphorus and potassium fertilizers are 64 (35.8 - 120.7), 17 (3.2 - 56.8) and 2453 (744.9 - 4227.1) Bq/kg, respectively. With respect to organic fertilizers under investigation, the average radioactivity of 226Ra, 232Th and 40K are 42, 10 and 333 BqKg−1, respectively. Radium equivalent activity is not exceed 370 Bq/kg, the maximum permissible limit for radiation dose for all present samples. Average values of the three natural radionuclides measured in the brands of fertilizers used in Saudi Arabia are within the range of values reported in several other countries. This study could be useful as baseline data for radiation exposure to fertilizers and their impact on human health.

INTRODUCTION

In order to reach high agricultural productivity, the present practice of replacing nutrients in soils and consequently supplying substances is done by the application of chemical fertilizers, mostly compounds commercially named NPK (nitrogen (N), phosphorus (P) and potassium (K)) and NPKs (sulfate based fertilizer), the phosphorus portion is taken from phosphate rocks, which contain enhanced concentration of natural radionuclides, [1,2]. So, fertilizers redistribute naturally occurring radionuclides at trace levels throughout the environment and become a source of radioactivity, this phenomenon may result in potential radiological risks owing to external exposure during resident time in the farms and internal exposure through ingestion of food grown on fertilizer soils, [3]. The use of different types of fertilizers in the agricultural sector for the purpose of enhancing crop yield has become very common nowadays, fertilizers are usually used in reclaiming the land and improving the properties of crops. Many types of fertilizers are used in Saudi Arabia like: Super Phosphate, Urea Sulfate, and Ammonium Nitrate Sulfate. Using phosphate fertilizers over a period of many years could eventually increase the radium and uranium content of the soil and consequently increase the radiation dose which would result in the corresponding increase of the dose and causes diseases for human body, [4]. Also, during handling, packing and transporting fertilizers, some workers can receive additional external exposure at dose rates up to 0.8 mGy·h−1 [5]. So it is important to measure natural radioactivity in fertilizers and by-products, because the high radioactive content may lead to significant exposure of miners, manufacturers and end users. Furthermore, such measurements provide basic data for the estimation of the amount of radioactivity spread on agricultural land along with fertilizers.

Worldwide, several studies have been carried out in the vicinities of phosphate fertilizer industries [4,6-8]. All the studies mentioned, and others, have assessed the radioactive pollution caused by production plants on the human population. The main pathways of human exposure have been published by [9].

The aim of the present study is to: 1) determine the activity concentrations of naturally occurring radionuclides in a famous fertilizer samples of different brands of NPK fertilizer and organic fertilizers, which are available samples in Saudi markets, 2) provide a useful information in the monitoring of environmental contamination by natural radioactivity.

MATERIALS AND METHOD

Forty five samples of 9 different brands were collected from the local market of Al-Taif City, 30 samples chemical and 15 samples an organic fertilizer. The code type and other information about these samples are summarized in Table 1. All the brands of fertilizer are solid. After collection, the samples were ground, pulverized to fine powdered material, heated in the oven at 80°C for 24 h to remove moisture and then sealed in plastic containers to avoid any possibility of out gassing of radon. The samples were left a side for a month to allow the attainment of ^{226}Ra - ^{222}Rn Radioactive equilibrium. The activity concentrations of ^{226}Ra, ^{232}Th and ^{40}K for all equilibrium samples were measured by a gamma ray spectrometry using a NaI (TI) detector 3 × 3 inch with a 1024- channel computer analyzer. The detector has a peak efficiency of 1.2 × 10^{-5} at 1332.5 keV Co-60 and an energy resolution (FWHM) of 7.5% for 662 keV, detector employed with adequate lead shielding which reduces the background radiation.

THEORETICAL CALCULATIONS

Calculation the Activity Concentrations

The specific activity of 226Ra was evaluated from gamma-ray lines of 214Bi at 609.3 keV and 214Pb at 351.9 keV, while the specific activity of 232Th was evaluated from gamma-ray lines of 212Pb at 238.6 keV, 228Ac at 911.1 keV. The specific activity of 40K was determined directly from its 1460.8 keV gamma-ray line. Activity calculations have been carried out using the procedure given by [10] the activity concentrations in each sample were evaluated using the following relation [11,12]:

$$A_S\left(\mathrm{Bq}\cdot\mathrm{kg}^{-1}\right)=C_a/\varepsilon P_r M_s \qquad (1)$$

where C_a is the net gamma counting rate (counts per second), ε the detector efficiency of the specific γ-ray, P_r the absolute transition probability of Gamma-decay and Ms the mass of the sample (kg).

Radium Equivalent Activity (Bq/kg)

To assess the real activity level of ^{226}Ra, ^{232}Th and ^{40}K in fertilizer, a common radiological index has been defined in terms of radium equivalent activity (Raeq) in Bq·kg^{-1} can be used, provides a very useful guideline in regulating the safety standards in radiation protection for a human population. The index was calculated through the following formula is based on the assumption that 370 Bq·kg^{-1} of ^{226}Ra, ^{259} Bq·kg^{-1} of ^{232}Th and ^{481} Bq·kg^{-1} of ^{40}K produce the same gamma-ray dose rate [13]:

$$Ra_{eq}\left(Bq\cdot kg^{-1}\right)=C_{Ra}+1.43C_{Th}+0.077C_{K} \qquad (2)$$

where C_{Ra}, C_{Th} and C_K are the specific activities (Bq/kg dry weight) of ^{226}Ra, ^{232}Th and ^{40}K, respectively.

Absorbed and Effective Dose

The measured activity of ^{226}Ra, ^{232}Th and ^{40}K were converted into doses (nGy·h−1 Bq^{-1}·kg^{-1}) by applying the factors 0.462, 0.604 and 0.0417 for radium, thorium and potassium, respectively, [14].

Table 1. Some information of the collected fertilizer samples

Sample code	Composition
F1	20N-20P-20K
F2	12N-10P-10K
F3	12N-12P + 2MgO + 6S
F4	14N-12P-14K
F5	15N-30P-15K
F6	20N-20P-20K
F7	Sheep fertilizer
F8	Cow fertilizer
F9	Natural organic

These factors were used to calculate the total absorbed gamma dose rate in air at 1 m above the ground level using the following equation:

$$\text{Absorbed dose } D\left(\text{nGy}\cdot\text{h}^{-1}\right) = 0.462C_{Ra} + 0.604C_{Th} + 0.0417C_{K} \quad (3)$$

Where C_{Ra}, C_{Th} and C_{K} are the activities (Bq·kg^{-1}) of radium, thorium and potassium in the samples. To estimate annual effective doses, must be taken into account of 1) the conversion coefficient from absorbed dose in air to effective dose and 2) the indoor occupancy factor. The annual effective doses are determined as follows [14]:

$$\text{Outdoor annual effective dose}(\text{mSv}) = D\left(\text{nGy}\cdot\text{h}^{-1}\right)\times 8760\ \text{h}\times 0.2\times 0.7\text{Sv}\cdot\text{Gy}^{-1}\times 10^{-6} \quad (4)$$

Annual estimated average effective dose equivalent received by a member is calculated using a conversion factor of 0.7 Sv·Gy^{-1}, which is used to convert the absorbed rate to annual effective dose with an outdoor occupancy of 20% [15].

RESULTS AND DISCUSSION

The measured concentrations of natural radionuclides ^{226}Ra, ^{232}Th and 40K in fertilizer samples are shown in Table 2. In chemical fertilizers, the concentration of ^{226}Ra varies from 35.8 to 120.7 Bq·kg^{-1} with an average value of 64 Bq·kg^{-1}, the concentration of ^{232}Th varies from 3.2 to 56.8 Bq·kg^{-1} with an average value of 17 Bq·kg^{-1}, whereas the concentration of ^{40}K exists in the range from 744.9 - 4227.1 Bq·kg−1 with an average value of 2453 Bq·kg^{-1}. For the organic fertilizer, the average activity of ^{226}Ra, ^{232}Th and ^{40}K are 42, 10 and 333 Bq·kg^{-1}, respectively. The results showed that the chemical fertilizers have the highest emitter of the radiation in comparison with the organic samples under study.

Table 2. The activity concentrations of ^{226}Ra, ^{232}Th and ^{40}K (Bq/Kg) in chemical and organic fertilizers.

Fertilizer type	Activity concentration (Bq/Kg)		
	^{226}Ra	^{232}Th	^{40}K
F1	(29.3 - 67.3) 45.7 ± 2	(8.58 - 27.58) 16.5 ± 7	(1167.3 - 227.3) 2373.2 ± 18
F2	(45.5 - 87.9) 63.4 ± 8	(6.5 - 11.12) 9.6 ± 0.5	(788.5 - 2502.9) 2010.5 ± 23
F3	(35.8 - 81.1) 62.9 ± 8	(11.51 - 56.81) 29.1 ± 3	(2138.1 - 4227. 2) 3076 ± 37
F4	(45.5 - 75) 58.6 ± 7	(8.35 - 19.48) 15.2 ± 0.4	(749.9 - 4127.1) 3065.7 ± 43
F5	(46.4 - 85.2) 66.4 ± 8	(12.21 - 37.72) 22.9 ± 2	(754.6 - 3175.8) 2155.2 ± 22
F6	(54.9 - 120.7) 83.9 ± 11	(3.16 - 14.47) 9.2 ± 0.4	(1048.8 - 2959.1) 2038.3 ± 35
Average Value	64	17	2453
F7	(21 - 45.3) 31.6 ± 1.6	(3.2 - 9.8) 6.3 ± 0.3	(115.2 - 396.3) 277.3 ± 12
F8	(25.8 - 48.6) 41.1 ± 1.7	(4.77 - 11.36) 7.5 ± 0.3	(135.4 - 523.8) 284.5 ± 8
F9	(45.7 - 70.3) 54.2 ± 1.3	(4.95 - 21.31) 16.9 ± 3	(135.4 - 523.8) 410.7 ± 14
Average value	42	10	333

Table 3, presents, the range value of radium equivalent activity (Raeq) for chemical fertilizers varied from 224.9 to 349.7 Bq $\cdot$kg^{-1} with an average value 275 Bq $\cdot$kg^{-1}, and for organic the average value 77 Bq $\cdot$kg^{-1}, therefor, the radium equivalent activity is not exceed 370 Bq/kg, the maximum permissible limit for radiation dose for all samples. So, the results of this study will contribute to the national data regarding natural radioactivity levels in fertilizers used in Saudi Arabia.

The absorbed and annual effective dose rates from the samples were calculated as shown in Table 3. The minimum and maximum values of absorbed dose and outdoor annual effective dose in the chemical samples studied in the present work, were found to vary from 118 nGy $\cdot h^{-1}$ to 155 nGy $\cdot h^{-1}$ with an average value 126 nGy/h, and 0.13 $mSvy^{-1}$ to 0.19 $mSvy^{-1}$ with an average value 0.15 $mSvy^{-1}$, respectively. For organic fertilizer the maximum value is 77.2 nGy $\cdot h^{-1}$ and the minimum is 40.8 nGy $\cdot h^{-1}$ with an average value is 57 nGy $\cdot h^{-1}$. The larger value given for the absorbed dose rate is of 55 nGy/h (24 - 85 nGy/h, [16]. A more recent UNSCEAR report gave wide range (18 - 93 nGy/h) with an average value of 59 nGy/h, [14]. The results showed that absorbed dose rate for the chemical fertilizer samples are higher than the world limits, but the effective dose rates for all samples were not exceed the recommended value 1 mSv. Therefore, all the samples analyzed in the present work satisfy the safety criterion for general public [17]. Hence, most of these samples do not pose much health hazard for the population.

Table 3. Radium equivalent, absorbed dose and dose equivalent for different samples

Sample of fertilizer	Ra_{eq} (Bq/Kg)	Absorbed dose ($nGy \cdot h^{-1}$)	Outdoor (mSv)
F1	257.5	118	0.15
F2	240.6	103.7	0.13
F3	349.7	155.3	0.19
F4	332.8	155.8	0.19
F5	245.2	118.7	0.15
F6	224.9	104.2	0.13
Average values (chemical fertilizers)	275	126	0.15
F7	40.8	28.6	0.03
F8	52.3	35.3	0.04
F9	77.2	51	0.06
Average values (organic fertilizers)	57	38	0.04

Figures 1 and 2, describe the average activity of 226Ra, 232Th, 40K (Bq $\cdot Kg^{-1}$) and the radium equivalent with absorbed dose of different fertilizer used in Saudi Arabia, respectively. Table 4 shows average concentrations of the radionuclides in our study compared with the reported values from other countries. It is found that for NPK fertilizers the average 226Ra concentration of (64 $Bqkg^{-1}$) is the lowest activity concentration values in comparison with other reported values concentrations except that of Finland (54 $Bq \cdot kg^{-1}$).

The average 232Th concentration of (17 $Bq \cdot kg^{-1}$) is found lower than the reported concentration from Egypt (67 $Bq \cdot kg^{-1}$), India (28 $Bq \cdot kg^{-1}$), USA (49 $Bq \cdot kg^{-1}$) and Brazil (80 $Bq \cdot kg^{-1}$), but it is higher than the reported concentrations from Saudi Arabia-Reyadh (23 $Bq \cdot kg^{-1}$), Finland (11 $Bq \cdot kg^{-1}$) and Germany (15 $Bq \cdot kg^{-1}$). The average ^{40}K concentration of (2453 $Bq \cdot kg^{-1}$) is lower than the values reported concentrations from Finland (3200 $Bq \cdot kg^{-1}$) and Nigeria (2729 $Bq \cdot kg^{-1}$) and higher than the other values under comparison.

CONCLUSION

Concentrations of 226Ra, 232Th and 40K in thirty five fertilizer samples collected from Al-Taif city (Saudi Arabia) have been measured using gamma spectrometry. Average concentrations of ^{226}Ra, ^{232}Th and ^{40}K are 64, 17 and 2453 $Bq \cdot kg^{-1}$, respectively. These values have been compared with the world wide reported data.

Table 4. Comparison of activity concentrations of 40K, 226Ra and 232Th in Saudi Arabian samples and other countries

Country	Sample	Activity Concentrations (Bq/Kg)			Ra_{eq} (Bq/Kg)	Reference
		^{226}Ra	^{232}Th	^{40}K		
Egypt	NPK	366	67	4	462	[18]
India	NPK	79	28	1024	198	[19]
Germany	NPK	520	15	720	597	[20]
Nigeria	NPK	143	9	2729	366	[21]
Brazil	NPK	420	80	153	546	[22]
Finland	NPK	54	11	3200	316	[23]
USA	NPK	780	49	200	865	[24]
Saudi Arabia (Riyadh)	NPK	75	23	2059	266	[6]
Saudi Arabia	NPK	64	17	2453	275	Present work

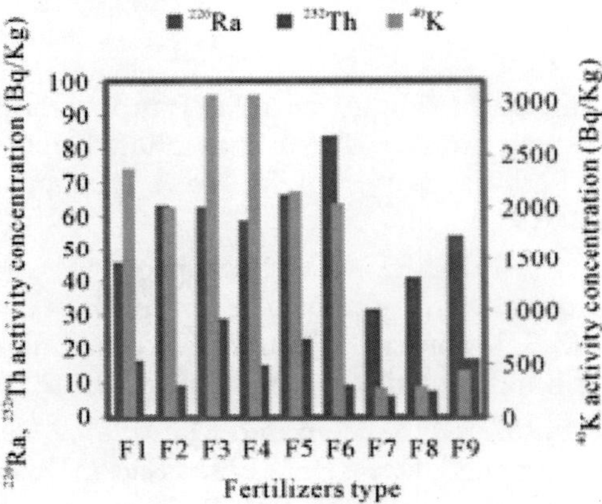

Figure 1. Average activity concentrations of ^{226}Ra, ^{232}Th and ^{40}K (Bq/Kg) of different fertilizer type from Saudi Arabia.

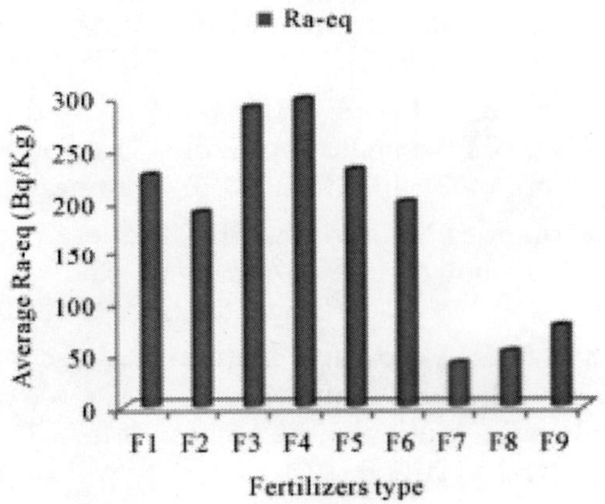

Figure 2. Average radium equivalent (Bq/Kg) in chemical and organic fertilizers used in Saudi Arabia.

The result showed that the highest values presented in chemical fertilizers and the lowest values were found in the organic fertilizers. Radium equivalent activity for all present samples were below the recommended limits (370 Bq $\cdot$kg^{-1}). The obtained results are in good agreement with those determined for Riyadh samples. From the calculated value it is recommended to use organic Fertilizer rather than chemicals.

REFERENCES

1. E. M. K. Ashraf, R. H. Higgy and M. Pimpl, "Radiologi-cal Impacts of Natural Radioactivity in Abu-Taror Phos-phate Deposits Egypt," Journal of Environmental Radio-activity, Vol. 55, No. 3, 2001, pp. 255-267. doi:10.1016/S0265-931X(00)00193-4
2. J. I. Skorovarov, L. I. Rusin, A. V. Lomonsov, H. Cha-forian, A. Hashemi and H. Novaseqhi, "Development of Uranium Extraction Technology from Phosphoric Acid Solutions with Extract," Proceeding of International Conference on Uranium Extraction from Soil, Vol. 217, 2000, pp. 106-113.
3. S. Rehman, N. Imtiaz, M. Faheem and Matiullah, "De-termination of 238U Contentsin Ore Samples Using CR-39 Based Radon Dosimeter Disequilibrium Case," Radia- tion Measurements, Vol. 41, 2006, pp. e471-e476.
4. S. Righi, P. Lucialli and L. Bruzzi, "Health and Environ- mental Impacts of a Fertilizer Plant Part I: Assessment of Radioactive Pollution," Journal of Environmental Radio- activity, Vol. 82, No. 2, 2005, pp. 167-182. doi:10.1016/j.jenvrad.2004.11.007
5. A. G. E. Abbady, M. A. M. Uosif and A. El-Taher, "Natural Radioactivity and Dose Assessment for Phos- phate Rocks from Wadi El-Mashash and El-Mahamid Mines, Egypt," Journal of Environmental Radioactivity, Vol. 84, No. 1, 2005, pp. 65-78. doi:10.1016/j.jenvrad.2005.04.003
6. E. M. Ashraf, Khater and H. A. AL-Sewaidan, "Radiation Exposure Due to Agricultural Uses of Phosphate Fertiliz-ers," Radiation Measurements, Vol. 43, No. 8, 2008, pp. 1402-1407. doi:10.1016/j.radmeas.2008.04.084
7. N. C. Silva, E. Fernandes, M. Cipriani and M. Taddei, "The Natural Radioactivity of Brazilian Phosphogypsum," Journal of Radioanalytical and Nuclear Chemistry, Vol. 249, No. 1, 2001, pp. 251-255. doi:10.1023/A:1013215215484
8. A. El-Taher and S. S. Althoyaib, "Natural Radioactivity Levels and Heavy Metals in Chemical and Organic Fer- tilizers Used in Kingdom of Saudi Arabia," Applied Ra- diation and Isotopes, Vol. 70, No. 1, 2012, pp. 290-295. doi:10.1016/j.apradiso.2011.08.010
9. M. Gascoyne, "Geochemistry of the Actinides and Their Daughters," In: M. Ivanovich and R. S. Harmon, Eds., Uranium Series Disequilibrium Applications to Earth, Marine and Environmental Science, Clarnandon Press, Oxford, 1992.
10. B. Y. Lalit and T. V. Ramachandran, "Natural Radiation Environment," 3rd Edition, Department of Energy, 1980, p. 800.
11. N. Ibrahim, "Determination of Natural Radioactivity in Fertilizers Using Gamma Ray," Spectroscopy Journal of Nuclear Physics and Chemistry, Vol. 51, No. 46, 1998, p. 621.
12. A. G. E. Abbady, "Estimation of Radiation Hazard Indi-ces from Sedimentary Rocks in Upper Egypt," Applied Radiation and Isotopes, Vol. 60, No. 1, 2004, pp. 111- 114. doi:10.1016/j.apradiso.2003.09.012
13. I. Beretka and P. I. Mathew, "Natural Radioactivity of Australian Building

Materials, Waste and By-Products," Health Physics, Vol. 48, No. 1, 1985, pp. 87-95. doi:10.1097/00004032-198501000-00007

14. UNSCEAR, "United Nations Science Committee on the Effects of Atomic Radiation Report to the General As- sembly, with Scientific Annexes," United Nations, New York, 2000.

15. UNSCEAR, "United Nations Scientific Committee on the Effect of Atomic Radiation, Report to the General As- sembly," United Nations, New York, 1993.

16. UNSCEAR, "Sources, Effects and Risks of Ionizing Ra- diation," United Nations, New York, 1988.

17. ICRP, "International Commission on Radiological Pro-tection, the 1990-1991 Recommendations of the Interna- tional Commission on Radiological Protection, Publica- tion 60," Pergamon, Oxford, 1991.

18. N. K. Ahmed and A. M. El-Arabi, "Natural Radioactivity in Farm Soil and Phosphate Fertilizer and Its Environ- mental Implication in Qena Governorate, Upper Egypt," Journal of Environmental Radioactivity, Vol. 84, No. 1, 2005, pp. 51-64. doi:10.1016/j.jenvrad.2005.04.007

19. P. Chauhn, R. P. Chauhan and M. Gupta, "Estimation of naturally Occurring in Radionuclides Fertilizers Using Gamma Spectrometry and Elemental by XRF and XRD Techniques," Microchemical Journal, 2012, in Press.

20. K. Khan, H. M. Khan, M. Tufail and N. Shmsf, "Gamma Spectrometric Studies of Single Super Phosphate Fertil-izer Samples," In: Proceedings of National Seminar on Occupational Safety in Mining & Industries, Peshawar, 1996.

21. N. N. Jibiri1 and K. P. Fasae, "Activity Concentrations of 226Ra, 232Th and 40K in Brands of Fertilizer Used in Nige- ria, Radiation Protection Dosimetry," Vol. 148, No. 1, 2012, pp. 132-137.

22. C. H. Saueia, B. P. Mazzilli and D. I. T. Favaro, "Natural Radioactivity in Posphogypsum and Fertilizer in Brazil," Journal of Radioanalytical and Nuclear Chemictry, Vol. 264, No. 2, 2005, pp. 445-448. doi:10.1007/s10967-005-0735-4

23. R. Mustonen, "Radioactivity of Fertilizer in Finland," Science of the Total Environment, Vol. 45, 1985, pp. 127- 134. doi:10.1016/0048-9697(85)90212-8

24. R. J. Gulmond and S. T. Windham, "Radionuclide's in the Environment," Technical Note No. ORP/CSD-75-3, USEPA, 1975, pp. 228-229.

Chapter 6

EFFECT OF ORGANIC FERTILIZERS USED IN SANDY SOIL ON THE GROWTH OF TOMATOES

Yongxia Hou[1], Xiaojun Hu[1*], Wenting Yan[2], Shuhong Zhang[2], Libin Niu[3]

[1]Shenyang University, Key Laboratory of Regional Environment and Eco-Remediation, Ministry of Education, Shenyang, China; houyongxia@126.com, *Corresponding Author: hu-xj@mail.tsinghua.edu.cn

[2]College of Horticulture, Shenyang Agricultural University , Shenyang, China; yanwenting2000@126.com, zhangsh024@163.com

[3]The Affiliated Hospital of Liaoning University of Traditional Chinese Medicine, Shenyang, China; xiaojun7770@163.com

ABSTRACT

In order to reveal the regulating capacity of organic fertilizers on sandy soil, pots experiments were carried out. The growth of tomatoes planted on sandy soil amended by organic fertilizers was measured. Organic fertilizers can be helpful to improve the plant height, stem diameter, the aerial parts fresh weight, root fresh weight, leaf photosynthetic rates and photosynthesis, and lay a good foundation for the growth of tomatoes. The effect of organic fertilizer is the most significant. Among all the treatments of adding 2.5%, 5%, 10% organic fertilizers, adding 10% organic fertilizers are the best. It can significantly enhance the growth and photosynthesis of tomatoes, and it is among the best of these three soil treatments for sandy soil.

INTRODUCTION

Sandification is an environmental problem for the world. China is one of the countries suffered desertification seriously. In 2011, the State Forestry Administration (SFA) has announced the results of forth national monitoring of desertification and sandification. The results show that, by the end of 2009, the area of national sandy lands, 18.03% of national lands, is 1, 731, 100 hm2. In these areas, 310,000 hm2 became sandy soil obviously. Some are still expanding. Sandy soil is a poor soil which has low contents of organic matters and nutrition. Owing to the loose texture and the gap between particles, the capacity of saving water and nutrients is poor. Not only was development soil productivity had constrained, but also the environmental deterioration and economic losses were getting worse and worse. Soil amendments have been reported to change sandy soil texture in order to increase water holding capacity [1]. Adding organic fertilizers in sandy soil, the growth of the plant growth was affected [2,3]. and in some cases neutralize soil acidity [3] and enhance soil catalase activity [4]. Horticultural production has an especially reliant relationship with soil. The overspread of cultivated lands is limited, but the amendable capacity is unlimited [5]. Soil amendments can not only change the characters of sandy soil, but enhance yield, quality [6] and stress resistance [7], promote the growth of horticultural plants. As the foundations of tomato growth, development and yield, photosynthesis is an important indicator in the fields of breeding, cultivation and environmental stress. The experiments reported here evaluated the use of organic fertilizers to improve the plant height, stem diameter, the aerial parts fresh weight, root fresh weight, leaf photosynthetic rates and photosynthesis by two kinds of sandy soil, for revealing the regulating capacity of organic fertilizers on sandy soil.

MATERIALS AND METHODS

Materials and Experimental Design

Experiments were conducted at the greenhouse of Shenyang University. The soil is sandy soil. Strong sandy soil was taken from

Dong Liujiazi located in 42° 20′ 43N, 122° 43′ 19E. Weak sandy soil were taken from Xiu Shuihe in Faku located in 42° 21′ 26N, 123° 00′ 21E. Organic fertilizer was purchased from flower market. The variety of tomato for experiment is Fuyou Dafen.

These were pots experiments. These soil amendments of organic fertilizers were respectively added to two kinds of sandy soil at the mass ratios of 2.5%, 5% and 10%. There were 20 treatments. Each one had 5 repetitions. The experimental treatments are showed as follows (**Table 1**).

Tomatoes were seeded on February 22, 2012 and planted in pots on March 28. Except the amounts of soil amendments and sandy soil, each one was used by one routine management.

Table 1. Treatments.

Treatments	Organic fertilizer ($g \cdot kg^{-1}$)	Strong sandy soil($g \cdot kg^{-1}$)	Weak sandy soil ($g \cdot kg^{-1}$)	Mass rotio (%)
CK(A)	0	1000	0	0
A1	25	975	0	2.5
A2	50	950	0	5
A3	100	900	0	10
CK(B)	0	1000	0	0
B1	25	0	975	2.5
B2	50	0	950	5
B3	100	0	900	10

Experimental Methods

The parameters of plant height, stem diameter, the aerial parts fresh weight, root fresh weight were investigated by conventional techniques. By using Lc Pro+ photosynthesis system, the light responses of photosynthesis in leaves of 60 days after planted tomatoes were studied under 2 different soil water conditions: drought (Relative Soil Water Content (RSWC) is 60%) and saturated-water (RSWC is 100%). The measure time was during 10:00 - 12:00 on sunny days. The third healthy functional leaves were chosen from the top as samples, every treatment took 5 samples. The average values were as the final measured results. Artificial light source were used to control Photosynthetic Photon Flux Density (PPFD). The PPFD were 600, 800, 1000, 1200, 1400, and 1600 μmol · $(m^2 \cdot s)^{-1}$. Determined time was 60 s under each PPFD.

RESULTS AND DISCUSSION

Effect of Adding Organic Fertilizers on the Growth of Tomato

Adding 2.5%, 5%, 10% organic fertilizers in strong sandy soil obviously improved the parameters of the plant height, stem diameter, the aerial parts fresh weight, root fresh weight of tomato leaves (**Table 2**). And it had the same trend in weak sandy soil. The parameters of the plant height, stem diameter, the aerial parts fresh weight, root fresh weight improve with the increasing of organic fertilizers.

Effect of Adding 2.5% Organic Fertilizers on Photosynthetic Rate of Tomato

Adding 2.5% organic fertilizers in strong sandy soil obviously improved photosynthetic rates of tomato leaves (**Figure 1**). Photosynthetic rates of plants treated by organic fertilizers increased rapidly with the raising of PPFD. With raising trends of the rates got slowly later. The results showed that when the RSWC was 60%, the photosynthetic rate of adding organic fertifizer increased apparently to the max value of 7.13 μmol · $(m^2 \cdot s)^{-1}$ under 1600 μmol · $(m^2 \cdot s)^{-1}$ PPFD. The increased tendency of tomato photosynthetic rates of every treatments under 100% RSWC was the same as that in drought condition (RSWC is 60%). The max was 8.74 μmol · $(m^2 \cdot s)^{-1}$ from organic fertifizer used treatment. By comparing the treatment of A1 and B1, 2.5% soil amendments made a better influence in weak sandy soil than that in strong sandy soil.

Effect of Adding 2.5% Organic Fertilizers on Photosynthetic Rate of Tomato

According to **Figure 2**, after adding 5% organic fertilizers in strong sandy soil, the photosynthetic rate of tomatoes got an obvious enhancement. With the PPFD

Table 2. The effect of organic fertilizers on the growth of tomato in flowering stage.

Treatments	plant height (mm)	stem diameter (mm)	the aerial parts fresh weight (g)	root fresh weight (g)
CK(A)	201.08	2.84	3.15	0.51
A1	207.45	2.9	3.24	0.58
A2	257.3	3.0	3.35	0.75
A3	286.54	3.1	4.11	0.85
CK(B)	204.92	2.85	3.54	0.65
B1	214.38	2.96	4.03	0.69
B2	223.19	3.14	5.09	0.71
B3	294.9	3.38	5.53	0.82

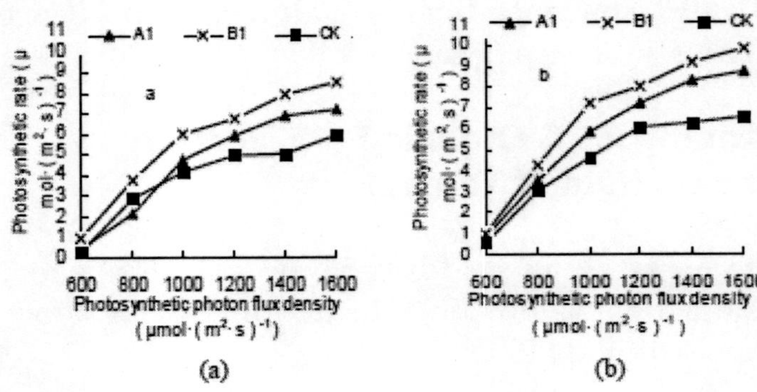

Figure 1. Photosynthetic rate-light response curves of tomato under adding 2.5% soil amendments in strong sandy soil and weak sandy soil: (a) Drought condition (RSWC is 60%); (b) Saturated-water condition (RSWC is 100%).

raising, the photosynthetic rates of the treatments in- creased rapidly. It was an evident improvement of tomato photosynthetic rate on sandy soil amended by organic fertilizers under 60% RSWC, while the PPFD was 1600 μmol · $(m^2 \cdot s)^{-1}$, to the max 7.69 μmol · $(m^2 \cdot s)^{-1}$. After rewatering, the tomato photosynthetic rates tendency of every treatments was simillar to that in drought condition (RSWC is 60%) . The max was 9.62 μmol · $(m^2 \cdot s)^{-1}$ from organic fertilizers used treatment. After applying 5% soil amendments on weak sandy soil, it showed that the tomato photosynthetic rate of each treatment in drought condition was higher than that in saturated-water condition by analyzing the effects of different relative soil water contents. The amending effect of adding 5% amendments on weak sandy soil was better than that on the strong one.

Effect of Adding 10% Soil Amendments on Photosynthetic Rate of Tomato

The tomato photosynthetic rate under 10% organic fertilizers used was increasing with photosynthetic pho- ton flux density (**Figure 3**). All got the top value when the PPFD was 1600 μmol · $(m^2 \cdot s)^{-1}$. The effect of the treatment of adding 10% organic fertilizers was better than that of ck in drought condition, the max 7.41 μmol $\cdot(m^2 \cdot s)^{-1}$. When RSWC was 100%, the max values of organic fertilizers was 9.21 μmol $\cdot(m^2 \cdot s)^{-1}$.10% organic fertilizers made a better influence in weak sandy soil than that in strong sandy soil. The effects of all the treatments on enhancing tomato photosynthetic rate were better than that of control sample.

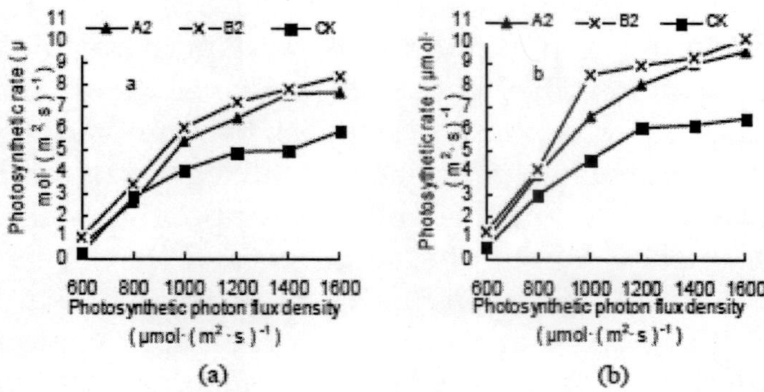

Figure 2. Photosynthetic rate-light response curves of tomato under adding 5% soilamendments in strong sandy soil and weak sandy soil.

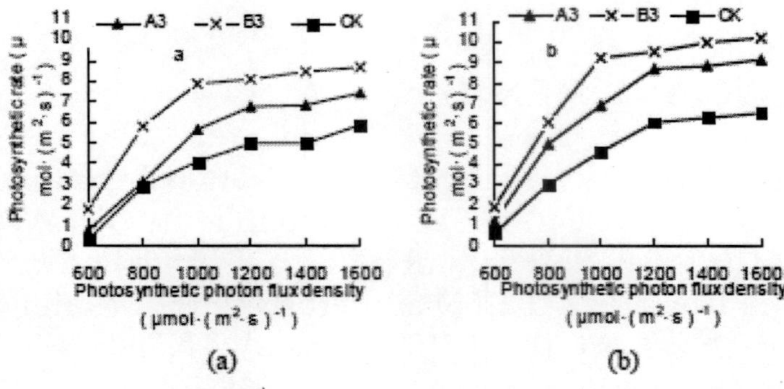

Figure 3. Photosynthetic rate-light response curves of tomato under adding 10% soil amendments in strong sandy soil and weak sandy soil: (a) Drought condition (RSWC is 60%); (b) Saturated-water condition (RSWC is 100%).

CONCLUSIONS

This study showed that after adding two kinds of sandy soil by adding 2.5%, 5% and 10% organic fertilizers under different relative soil water contents, the growth of the plant height, stem diameter, the aerial parts fresh weight, root fresh weight and the photosynthetic rates of tomato affected by organic fertilizer were all enhanced. the effect of using organic fertilizers was obvious. It showed that organic fertilizers could enhance crop physiological activity effectively, improve photosynthesis, and promote the growth of tomato. In addition, they could also enhance the accumulation of photosynthetic production, and establish the foundation of yield and quality. It is probably that organic fertilizers have multi-nutrients for the crop growth, and the ability of improving soil texture and nutritional status. Under the same organic fertilizers used, the effect in saturated-water condition was better than that in drought condition. It showed that the photosynthesis of tomato rebounded after rewatering. The reason is probable that organic fertilizers could enhance the soil water-nutrition holding capacity, and amend the weakness of sandy soil. By comparing the effects on different sandy soil, the photosynthetic rates on weak soil all higher than that on strong soil. The general analysis showed that adding 10% organic fertilizers was the best way to improve photosynthetic rate and growth of tomato.

ACKNOWLEDGEMENTS

This work has been supported by National Natural Science Foundation of China (21277093), Liaoning BaiQianWan Talents Program (2010921004), Natural Science Foundation of Liaoning Province of China (201102156) and Program for Liaoning Excellent Talents in University (LR2011034) and the National Science and Technology Supporting Project (2011BAJ06B02).

REFERENCES

1. Busscher, W.J., Novak, J.M. and Caesar-tonthat, T.C. (2007) Organic matter and polyacrylamide amendment of Norfolk loamy sand. *Soil & Tillage Research*, 171-178.
2. Zhang, B.B., Guo, J.B. and Jiang, K.Y. (2011) Effects of Arkadolith soil modifier on sand soil's properties and growth of Astragalus Mongolicum. *Bulletin of Soil and Water Conservation*, 190-194.
3. Cui, J.Y., Li, Y.L. and Su, Y.Z. (2002) Experiment of using fermented waste

residue from alcohol production to improve sandy soil. *Journal of Desert Research*, 368-371.

4. Cui, L.L., Li J.J. and Zou, G.Y. (2004) Effect of bentonite
5. on sandy soil fertility. *Acta Agriculture Boreali- Sinica*, 76-80.
6. Zhang, Z.X. (2005) Theory and practice of half dry identify water saving agriculture on northeastern China. China Agriculture Press.
7. Chen, F.S., Zeng D.H. and Chen, G.S. (2003) Effects of peat and weathered coal on physiological characteristics
8. and growth of Chinese cabbage on aeolian sandy land. *Journal of Soil and Water Conservation*, 152-155.
9. Giri, B., Kapoor, R. and Mukerjik, G. (2003) Influence of arbuscular mycorrhizal fungi and salinity on growth, biomass and mineral nutrition of Acacia auriculiformis. *Biology and Fertility of Soils*, **38**, 170-175. doi:10.1007/s00374-003-0636-z

Chapter 7

COMPLEMENTARY EFFECTS OF ORGANIC AND MINERAL FERTILIZERS ON MAIZE PRODUCTION IN THE SMALLHOLDER FARMS OF MERU SOUTH DISTRICT, KENYA

Edwin Mwiti Mutegi[1], James Biu Kung'u[2], Mucheru-Muna[2], Pypers Pieter[1], Daniel Njiru Mugendi[2]

[1]Tropical Soil Biology and Fertility (TSBF), Institute of CIAT, Nairobi, Kenya;

[2]Department of Environmental Sciences, Kenyatta University, Nairobi, Kenya

ABSTRACT

Low soil fertility is a major constraint to maize production in the small holder farms of Meru South District. This is mainly attributed to the mining of nutrients due to cropping without external addition of adequate nutrients. Mineral fertilizers are expensive hence unaffordable by most small holder farmers. The use of organic matter to increase and maintain soil fertility is being considered as a solution to help the low- income small holder farmers. A study was con- ducted in Mucwa location, Meru South District to determine the levels of complementarity between organic and mineral N amendments on maize yields and their influence on soil chemical properties. The experiment was set in a complete randomized block design (CRBD) with three replicates. The treatments were compared with the response obtained from control. The general soil fertility parameters changed slightly with Calcium, Magnesium and Potassium increasing in all treatments. The organic Carbon and total Nitrogen was higher in treatments that received sole organic N sources than in sole mineral N and a combination of organic and mineral N sources. The highest maize grain yield of 4.8 t $\cdot$ha^{-1} and 4.2 t $\cdot$ha^{-1} were realized from sole application of calliandra during the 2005 Short rains and 2006 Long rains cropping seasons. Generally the maize grain yields were lower in treatments with mineral N alone compared to the treatments with organics. Treatments with sole calliandra and sole tithonia had the highest benefit cost ratio (BCR), followed closely by manure treatment. More so, integration of organic and mineral N sources resulted to higher net benefit and BCR than the application of the re- commended rate of mineral fertilizers. Results obtained indicated that the use of either organic or combined organic/mineral N soil amendment appear to be superior to using mineral amendment sources alone.

INTRODUCTION

Soil fertility depletion in the smallholder farms is the fundamental biophysical root cause for declining per capita food production in sub-Saharan Africa. The soils in the central highlands of Kenya are Humic Nitisols with moderate to high inherent fertility [1]. However, the area has suffered gross soil nutrient mining due to continuous cropping coupled with low levels of nutrient inputs and poor nutrient conservation practices. The situation is further accentuated by mounting population growth and land scarcity [2, 3]. The results of this loss in soil productivity has been a continuous decline of maize yields in farmers' fields (to less than 2.0 t $\cdot$ha^{-1}) whilst the maize cultivars grown have a potential of greater than 6.0 t $\cdot$ha^{-1} [4,5]. The use of mineral fertilizers on staple food crops of maize (Zea mays L.) and beans (Phaseolus vulgaris L.) has generally been restricted to only a few farmers endowed with resources, such as cattle and land [6] and with high off-farm income [7]. The majority of the smallholder farmers, on the other hand have lacked the financial resources to purchase sufficient mineral fertilizers to replace the soil nutrients exported with harvested crop products. The situation is further aggravated by the fact that even the farmers using the inorganic inputs hardly use the recommended rates (60 kg ·N ·ha^{-1}) with most of them applying less than 20 kg ·N ·ha−1 [8].

Organic inputs are often proposed as alternatives to mineral fertilizer. For instance, [9] reported that soil in corporation of calliandra and Leucaena green biomass with or without fertilizer increased soil nitrogen by 1% -8% over a period of 4 years. During the same period, the total soil nitrogen declined by 2% - 4% when biomass was not applied. [10] reported that combination of mineral fertilizers with organic nutrient sources can be considered as better options for increasing fertilizer use efficiency as a result of improved synchronization of nutrient release and uptake by the crop [11]. A judicious combination of organic and mineral sources of nutrients may therefore be envisaged as it addresses both the problem of insufficient fertilizer supply and the large amounts of organic material required for nutrients supply.

A study was conducted in Mucwa sub location, Meru South District with the main objective of determining the levels of complementarities between organic and mineral N amendments on their influence on soil chemical properties and maize grain yields.

MATERIALS AND METHOD

Site Description

The study was conducted in Mucwa sub location, Chuka Division, in Meru South District. According to [1] the area is in upper midland 2 and 3 (UM2 and UM3) with an altitude of approximately 1373 m above sea level. This is a predominantly maize growing zone in the central highlands of Kenya. It experiences bimodal rains which range from 1200 to 1400 mm and mean temperature of about 200°C annually. The long rains (LR) are from March to June, and the short rains are from October to December. The predominant soil types are humicnitisols, commonly called the red Kikuyu loams. They are deep, well weathered, and free draining with a friable clay texture and moderate to high inherent fertility [1].

Experimental Treatments and Design

The experiment was laid out as a randomized complete block design (RCBD) with 3 replicates. The plots were measuring 6 × 4.5 m with 1 m and 1.5 m between and within plots respectively. The test crop was maize (Zea mays L., var. H513) planted at a spacing of 0.75 and 0.5 m inter- and intra-row, respectively. Three (3) seeds were sown per hole and thinned four weeks later to 2 plants. Nine external soil fertility amendment inputs were applied to give an equivalent amount of the recommended rate of nitrogen to meet maize nutrient requirements for an optimum crop production in the area (60 kg ·N ·ha^{-1}).

The tenth treatment was absolute control (no soil fertility enhancement input) representing farmers on the lower end of resource endowment. The organic inputs (biomass transfer) were harvested, weighed, chopped and incorporated into the soil to a depth of 15 cm during land preparation. Calcium Ammonium Nitrate (CAN) was the source of mineral N and at all its application rates, one-third was applied 4 weeks after planting (WAP) and the other two-thirds was applied 8 WAP. To prevent phosphorous deficiencies confounding N response, all plots received P application at 60 kg ·P ·ha^{-1}. The average nutrient composition of the organic inputs that were incorporated in the two seasons is shown in Table 1.

Sampling and Analysis

Soil samples were collected from each plot at the beginning of the 2004 Short rains (SR) and the end of the experiment (2006 LR). The sampling depth will be 0 - 15 cm. Six samplings were taken randomly in each plot using an alderman auger and then bulked to one sample to eliminate variability. A sub sample was taken for analysis of total N, C and pH and exchangeable bases (K, Ca and Mg) using standard methods. At maturity, maize was harvested and the fresh weight of both grain and stover taken. The maize was then air-dried and the dry weight taken and expressed on a 12.5% water content basis. Treatment effects on maize yields and soil chemical properties were subjected to analysis of variance (ANOVA) using Genstat programme. Treatment means found to be significantly different from each other were separated by least significant differences (LSD) at $p < 0.05$.

Economic Analysis

The economic returns from the application of each treatment were calculated taking into account the total sales and total costs. The information used for benefit- cost analysis was collected at the specific time of each activity in the course of each season. The benefit-cost analysis was done using farm gate prices of the various inputs, however, all the organic amendments (except manure) did not have market prices in the area and were therefore costed in terms of labor involved in harvesting and incorporation (Table 2).

Total cost included all the expenses for buying and applying the mineral fertilizers and the labor for collection, transporting and application of organic resources since they were collected near the experimental plots. The price of maize was taken to have a market value of USD 0.17 kg−1. Maize stover was used to feed cattle in the area (with a market value of USD 23 ton−1). To test for significant probability differences of the various organic and mineral fertilizers, ANOVA was carried out. The total cost was subtracted from the total sales to get the net benefit for the specific season. The benefit cost ratio was calculated by dividing the net benefits with the total cost.

Table 1. Average nutrient composition (%) of organic materials applied in the soil during the 2005 short and 2006 long rains at Mucwa, Meru South District, Kenya

Treatment	% N	% P	% K	% Mg	% Ca	Ash
Tithonia	3.0	0.2	2.0	0.7	1.6	13.2
Manure	1.0	0.3	0.9	0.4	1.4	46.1
Calliandra	3.3	0.2	1.1	0.4	0.9	5.8
SED	**0.72**	**0.03**	**0.34**	**0.10**	**0.21**	**12.39**

RESULTS AND DISCUSSIONS

Maize Grain Yields

One of the objectives of this study was to determine the effects of organic and mineral N sources on maize grain yields. Table 3 shows grain yields obtained during the two seasons (2005 SR and 2006 LR).

During the first season (2005 SR), sole calliandra treatment gave the highest maize grain yield of 4.8 t·ha^{-1}, followed by tithonia + 30 kg·N·ha^{-1} and 90 kg·N·ha^{-1} (3.6 t·ha^{-1}). The control gave the lowest maize grain yield across the treatments with 1.6 t·ha^{-1} followed closely by fertilizer at 30 kg·N·ha^{-1} (1.8 t·ha^{-1}). The yields obtained from fertilizer at 60 kg·N·ha^{-1} were not significantly different from the 90 kg·N·ha^{-1} (3.0 t·ha^{-1} and 3.6 t·ha^{-1} respectively).

The results obtained during the 2006 LR season revealed that maize grain yield was between 0.4 t·ha^{-1} and 4.2 t·ha^{-1}, which was against the expected grain yield of greater than 6 t·ha^{-1} (Var. H 513) for the area. During this season, calliandra treatment gave the highest grain yield of 4.2 t·ha^{-1}, while the control and application of 30 kg·N·ha^{-1} gave the lowest maize grain yield (0.4 t·ha^{-1}). Sole tithonia and tithonia + 30 kg·N·ha^{-1} gave significantly higher yields than all the other treatments (except in sole calliandra treatment). The yields obtained from application of 60 kg·N·ha^{-1} were not significantly different from those obtained in 90 kg·N·ha^{-1} treatment. Sole organics and integration of sole organics with 30 kg·N·ha^{-1} had higher yields than the recommended rate of mineral fertilizer (60 kg·N·ha^{-1}). The application of organic alone or in combination with mineral fertilizers led to increased maize yield compared to the control.

The higher maize grain yields obtained during the 2005 SR season from mineral fertilizer at 90 kg·N·ha^{-1} and 60 kg·N·ha^{-1} (though not significantly different from sole tithonia, sole manure and Tithonia + 30 kg·N·ha^{-1}) could be attributed to nutrients being readily available from the mineral fertilizers as compared to nutrients from organic residues which must first undergo decomposition before they are available for crop uptake. The split application of mineral N could have also resulted to minimal leaching losses and better synchrony of nutrient availability to maize crop demand. [12] suggested that split N application should be implemented so as to in- crease plant N uptake and decrease potential for N losses. Another aspect that contributed to high maize yields during the 2005 SR season was the even distribution of rainfall throughout during the first three months of the cropping season.

Table 2. Parameters used to calculate the economic returns for the different nutrient replenishment technologies at Mucwa, Meru South District, Kenya.

Parameter	Actual values
Price of TSP (46% N)	0.61 USD·kg^{-1}
Price of CAN (26% N)	0.77 USD·kg^{-1}
Labor cost	0.13 USD·h^{-1}
Labor cost for planting maize	10.5 USD·ha^{-1}
Labor for applying fertilizer	0.74 USD·h^{-1}
Labor for applying organic inputs	2.9 USD 100 kg^{-1}
Price of Maize	0.17 USD·kg^{-1}
Price of stover	0.023 USD·kg^{-1}

DM = Dry matter; Exchange rate 65 Ksh = USD 1 (November, 2007).

Table 3. Maize grain yield (t ·ha^{-1}) during the 2005 SR and 2006 LR at Mucwa, Meru South District, Kenya

Treatment	2005 SR	2006 LR
Calliandra	4.8^{a}	4.2^{a}
Tithonia	3.1^{bcd}	3.4^{b}
Manure	3.5^{bc}	2.4^{d}
Tithonia + 30 kg·N·ha^{-1}	3.6^{b}	3.2^{bc}
Calliandra + 30 kg·N·ha^{-1}	2.9^{d}	2.5^{cd}
Manure + 30 kg·N·ha^{-1}	2.9^{d}	3.0^{b}
Fertilizer (90 kg·N·ha^{-1})	3.6^{b}	2.3^{c}
Fertilizer (60 kg·N·ha^{-1})	3.0^{bc}	2.0^{de}
Fertilizer 30 kg·N·ha^{-1})	1.8^{e}	0.4^{f}
Control	1.6^{e}	0.4^{f}

Means with same letter in each column are not statistically different at $p < 0.05$.

This promoted rapid growth since the soil moisture deficits were eliminated. The lower yield obtained from fertilizer at 30 kg ·N ·ha^{-1} in comparison to the 60 kg ·N ·ha^{-1} is probably because the N supplied was not enough to meet the maize crop demand since the recommended mineral N application rate for the area is 60 kg ·N ·ha^{-1} [13].

Lower moisture regimes characterized the 2006 long rains cropping season with 79% of the rainfall being received within the first 40 days of the season. This may in part have been responsible for the suppressed performance of maize crop during the 2006 LR period. The low moisture regimes in the soil could also have meant that most of the organic materials did not fully decompose in time, thus N was not fully released in time, and if it was, water was not available for the mineralized nutrients to be taken up by the crop. Soil moisture content influences N mineralization and availability and subsequent maize growth and uptake [14, 15] noted that variability in climatic factors such as rainfall and temperature make the synchrony between nutrient release from tree litter and crop uptake an elusive goal to achieve in practical terms. Insufficient moisture has also been reported to limit the response of crops to nutrients [16]. During this season, the better performance in sole tithonia and tithonia + 30 kg ·N ·ha^{-1} is possibly due to high release of N through mineralization and this synchronized to plant uptake [17]. [18] noted that the overall secondary compounds (lignins and polyphenols) in tithonia are low compared with foliage of many trees. Tithonia contains 80% water that further contributes to rapid decomposition [19].

The relatively better performance from sole organics and integration of sole organics with 30 kg ·N ·ha^{-1} in comparison to sole mineral N sources could be due to provision of additional benefits (besides N) by the organic inputs to the soil chemical and physical properties that in turn influence nutrient acquisition and plant growth [11]. Principal among these is the soil moisture holding capacity and provision of other macro-nutrients like calcium and magnesium [20, 21]. Higher maize yields with organic and/or a combination of organics with mineral fertilizer has been reported elsewhere. For instance, re- search work by [22] and [23] have demonstrated that higher yields can be obtained when sole organics or their combinations with mineral fertilizer have been incorporated in comparison to sole mineral fertilizer treatment. [24,25] reported that a combination of organic and mineral nutrient sources has been shown to result into synergy and improved synchronization of nutrient release and uptake by plants leading to higher yields. More so, addition of green manure and animal waste helps to re-duce the total concentration of Al in the soils and thus reduce Al phototoxicity and increase crop growth [26, 27]. Another likely cause for the observed higher yields in the organic and/or mixed treatments was reduced water stress compared with sole mineral fertilizer treatments due to the presence of organic materials. The organic residues improve water holding capacity and moisture retention.

Table 4. Soil chemical properties (0 - 15 cm) at the beginning of the experiment (2004 SR) in Mucwa, Meru South District, Kenya

Treatment	pH	Ca	Mg	K	C%	N%
	(H_2O)	Exchangeable (cmol/kg)				
Manure	4.9^{cd}	0.9^{e}	0.18^{a}	0.3^{c}	1.9^{d}	0.23^{c}
Manure + 30 kg·N·ha^{-1}	5.2^{a}	1.5a	0.18^{a}	0.4^{b}	2.2^{ab}	0.25^{ab}
Tithonia	5.1^{ab}	$1.2b^{c}$	0.22^{a}	0.5^{a}	2.3^{a}	0.25^{ab}
Calliandra	5.0^{bc}	1.1^{cd}	0.18^{a}	0.4^{b}	1.9^{d}	0.24^{bc}
Tithonia + 30 kg·N·ha^{-1}	4.8^{d}	1.0^{de}	0.20^{a}	0.3^{c}	2.0^{cd}	0.26^{a}
Calliandra + 30 kg·N·ha^{-1}	5.0^{bc}	1.0^{de}	0.17^{a}	0.3^{c}	2.0^{cd}	0.23^{c}
Fertilizer (30 kg·N·ha^{-1})	4.9^{cd}	1.0^{de}	0.19^{a}	0.3^{c}	2.1^{bc}	0.24^{bc}
Fertilizer (60 kg·N·ha^{-1})	5.1^{ab}	1.3^{b}	0.23^{a}	0.3^{c}	2.1^{bc}	0.24^{bc}
Fertilizer (90 kg·N·ha^{-1})	4.9^{cd}	1.0^{de}	0.18^{a}	0.3^{c}	2.0^{cd}	0.25^{ab}
Control	4.8^{d}	1.0^{de}	0.20^{a}	0.4^{b}	1.9^{d}	0.24^{bc}

Means with same letter in each column are not statistically different at $p < 0.05$.

Soil Chemical Characteristics

The second objective of the study was to determine the influence of the various organic and mineral N resources on soil chemical properties. The general soil fertility parameters changed with the application of the various organic and mineral N sources as shown in Tables 4 and 5.

Table 5. Soil chemical properties (0 - 15 cm) at the end of 2006 Long rain season at the experimental site, Mucwa, Meru South Dis-trict, Kenya

Treatment	pH (H_2O)	Ca	Mg	K	C%	N%
		Exchangeable (cmol/kg)				
Manure	5.6^{a}	2.9^{b}	1.5^{a}	1.5^{a}	1.7^{bc}	0.28^{a}
Manure + 30 kg·N·ha^{-1}	5.2^{b}	3.4^{a}	1.1^{b}	1.3^{a}	2.2^{ab}	0.28^{a}
Tithonia	5.1^{bc}	2.3^{c}	0.8^{cd}	1.3^{a}	1.4^{c}	0.28^{a}
Calliandra	4.9^{de}	1.9^{d}	0.7^{d}	0.5^{bc}	2.5^{a}	0.27^{ab}
Tithonia + 30 kg·N·ha^{-1}	5.1^{bc}	1.3^{e}	0.6^{d}	0.3^{c}	2.0^{ab}	0.22^{d}
Calliandra + 30 kg·N·ha^{-1}	4.8^{e}	1.9^{d}	1.0^{bc}	0.4^{c}	1.5^{bc}	0.18^{e}
30 kg·N·ha^{-1}	5.1^{bc}	2.0^{d}	1.0^{bc}	0.5^{bc}	1.7^{bc}	0.22^{d}
60 kg·N·ha^{-1}	4.9^{de}	2.2^{cd}	0.7^{d}	0.5^{bc}	2.0^{ab}	0.26^{bc}
90 kg·N·ha^{-1}	5.0^{cd}	1.9^{d}	0.6^{d}	0.4^{c}	1.2^{cd}	0.26^{bc}
Control	4.8^{e}	1.9^{d}	0.7^{e}	0.5^{bc}	1.6^{bc}	0.25^{c}

Means with same letter in each column are not statistically different at $p < 0.05$.

The soil pH increased significantly in sole manure treatment ($p < 0.05$) while it declined (though not significantly) in treatments that received fertilizer at 60 kg·N·ha^{-1} and calliandra combined with half recommended rate of fertilizer at the end of the study period. However, the pH remained constant in manure + 30 kg·N·ha^{-1}, sole tithonia and the control treatments.

The pH of the soils ranged from 4.8 to 5.6 indicating that these soils were acidic. Under such conditions, the availability of the base forming cations is limited since the soil solution is mostly occupied by aluminium and hydrogen ions. The increment of soil pH with additions of manure could be attributed to the reduction of ex- changeable aluminium in the acidic soils. This reduction is considered to occur through aluminium precipitation or chelation of organic colloids [27]. Increased pH with manure additions could also be attributed to increased levels of exchangeable bases (K, Mg and Ca). The significant increase in pH with manure application corresponds with the findings by [28] and [29]. Increasing the pH of acidic soils improves plant-availability of macro- nutrients while reducing the solubility of elements such as Al and Mn [23, 30]. The magnitude of the rise in soil pH varies depending on the type of manure, its rate of application and the buffering capacity of the soil [31]. [32] noted that manures have the advantage of supplying essential plant elements either directly or indirectly by alleviating aluminium toxicity or producing organic ac- ids thereby increasing nutrient availability.

Exchangeable Ca increased significantly ($p < 0.05$) in all treatments with sole manure, manure + 30 kg ·N ·ha^{-1}, sole calliandra and calliandra + 30 kg ·N ·ha^{-1} treatments having the highest increase in Ca content. Exchangeable Mg also increased significantly in all treatments with sole manure having the highest Mg increment. Exchangeable potassium also increased significantly ($p < 0.05$) in sole manure, manure + 30 kg ·N ·ha−1 and sole tithonia treatments. The observed increase in exchangeable cations is consistent with the work of [33, 34] who reported an in-crease in exchangeable K and Ca after application of organic inputs. These are highly leached soils [1] and in such soils the order of adsorption of bases is Ca > Mg > K. Thus, continued leaching more of potassium followed by magnesium ions leaving relatively more calcium still adsorbed. In general, these soils were moderately acidic with medium to low nutrient stocks of exchangeable bases with a consequent effect on the availability of macro plant nutrients. This is partly because of the decomposition of organic materials yielding acids with release of hydrogen ions and also the reduced levels of bases due to leaching. Similar observations were made by [13] in some humic Nitisols in Kandara (Murang'a District, Kenya).

Organic carbon declined in all treatments except in sole calliandra treatment where it increased significantly by the end of the study period. The decreased C levels could be associated with the rapid decomposition rates in the study area and the fact that the organic input (tithonia) was of high quality. The N content in the organic materials also influence decomposition and N release. [11] noted that the N concentration in tithonia is higher than the critical levels of 2.0% and 2.5% below which net immobilization of N would be expected. However, in- crease in C levels in sole calliandra treatment could be attributed to lower rates of decomposition and mineralization of the biomass as a result of high polyphenol contents which bind with N and hence lower its decomposition rate and N release [35, 36]. On average, the organic carbon was higher in soils receiving organic amendments or a combination of mineral fertilizers with organic amendments compared to soils receiving mineral fertilizers alone by the end of 2006 LR cropping season. This was because whereas the organic material had a major impact on mineralization rate by increasing soil C directly, the effect of mineral N fertilizer was less pronounced since it increased C only indirectly by improving plant growth [37]. These results corroborates with those of [35] who reported that the use of organic amendments either singly or in combination with mineral fertilizers play a vital role in sequestering C and building up soil fertility.

Total N increased in all treatments except in tithonia + 30 kg ·N ·ha^{-1}, calliandra + 30 kg ·N^{-1} and mineral fertilizer (90 kg ·N ·ha^{-1}) which recorded an insignificant de- crease. On average, treatments that received sole organic N amendments were higher in total N than the treatments that received integrations (of organic and mineral fertilizers) and sole application of mineral N fertilizers. The relative increase of total N at the end of the experiment could be attributed to minimal uptake of total N due to poorly distributed rainfall in 2006 LR season (Data not shown). The poor distribution of rainfall resulted to reduced nutrient availability to the maize crop. [35] noted that even distribution of rainfall and availability of sufficient amounts of rainfall are very important factors to crop production without which crop yields decline or even fail. However, total N was highest in all treatments that received organic residues in comparison to sole mineral N treatments. This could be attributed to the fact that the organics must first of all undergo microbial de- composition unlike the mineral N fertilizers which are applied in plant available form with subsequent losses through leaching and denitrification very early in the season.

Cost-Benefit Analysis

An economic analysis was done on the different organic and mineral N inputs during the 2005 SR and 2006 LR season. The organic amendments (except manure) did not have market prices in the area and were thus costed in terms of labor involved in harvesting and in- corporation. The results are presented in Table 6.

Cost-benefit analysis (CBA) during the 2005 SR sea- son indicated sole calliandra treatment gave the highest net benefit of USD 1089 and was followed by tithonia + 30 kg ·N ·ha^{-1} (USD 844) which was not significantly different ($p < 0.05$) from sole manure and sole tithonia treatment (USD 800 and USD 621 respectively). The lowest net benefit was recorded in control treatment (USD 376) and was followed by 30 kg ·N ·ha^{-1} treatment with USD 488. The net benefits observed in 60 kg ·N ·ha^{-1} and 90 kg ·N ·ha^{-1} were not significantly different from each other (707 and USD 733 respectively). This indicates that the additional mineral N in the 90 kg ·ha^{-1} treatment did not offer an economic benefit since the observed maize grain yield in the 60 kg ·ha^{-1} and 90 kg ·ha^{-1} treatment (Table 6) was also not significantly different.

For the same period (2005 SR) sole calliandra yielded the highest benefit-cost ratio (BCR) of USD 12.4 and was followed by sole tithonia (USD 10.1) and sole manure treatment (USD 9.4). The control treatment had significantly higher BCR than 90 kg ·N ·ha−1 (USD 6.3 compared to USD 3.9 respectively). Treatments involving sole organics and/or their integrations with mineral N sources recorded a higher BCR during this season.

During the 2006 LR season, a similar trend was observed with sole calliandra giving the highest net benefit of USD 991 while the control and 30 kg ·N ·ha^{-1} recorded the lowest (USD 182 and USD 261 respectively). The observed BCR was also in the order of sole organic treatments > organics + mineral N treatments > sole mineral N treatments.

Table 6. Cost-benefit ratios for different soil fertility amendments on maize yields during the 2005 SR and 2006 LR at Mucwa, Kenya

Trt	Total sales (USD))		Total cost (USD)		Net benefit (USD)		Benefit-Cost ratio	
	SR 05	LR 06	SR 05	LR 06	SR 05	LR 06	SR 05	LR 06
1	1176^{a}	3342^{a}	88^{cd}	83^{cd}	1089^{a}	991^{a}	12.4^{a}	11.9^{a}
2	864^{bc}	2751^{b}	78^{de}	80^{cd}	786^{bcd}	834^{ab}	10.1^{b}	10.4^{a}
3	884^{bc}	1993^{d}	85^{cd}	76^{d}	799^{bc}	621^{cd}	9.4^{bc}	8.1^{bc}
4	949^{b}	2609^{bc}	105^{bc}	102^{bc}	844^{b}	779^{bc}	8.0^{cd}	7.6^{c}
5	764^{c}	2046^{cd}	95^{cd}	92^{bcd}	670^{cd}	605^{d}	7.1^{def}	6.6^{cd}
6	765^{c}	1195^{e}	98^{cd}	86^{cd}	667^{d}	407^{e}	6.8^{def}	4.7^{def}
7	922^{b}	1943^{d}	189^{a}	179^{a}	733^{bcd}	522^{d}	3.9^{g}	2.9^{f}
8	824^{bc}	1734^{d}	117^{b}	109^{b}	707^{cd}	544^{d}	6.0^{f}	5.0^{de}
9	573^{d}	551^{f}	84^{cd}	74^{d}	488^{e}	261^{f}	5.8^{f}	3.5^{ef}
10	436^{e}	448^{f}	60^{e}	50^{e}	377^{e}	182^{f}	6.3^{ef}	3.6^{ef}

Treatment (1 = Calliandra; 2 = Tithonia; 3 = Manure; 4 = Tithonia + 30 kg ·N ·ha^{-1}; 5 = Calliandra + 30 kg ·N ·ha^{-1}; 6 = Manure + 30 kg ·N ·ha^{-1}; 7 = 90 g ·N ·ha^{-1}; 8 = 60 kg ·N ·ha^{-1}; 9 = 30 kg ·N ·ha^{-1}; 10 = Control.

On average, integrations of organics and mineral fertilizers had a higher net benefit and BCR than sole application of mineral fertilizers. This indicates that sole organics or their integrations with mineral fertilizers is a more economically profitable investment amongst the small holder farmers in Meru South District.

Studies indicate that combination of organic and mineral nutrient sources ensure greater synchrony between nutrients released and plant uptake and hence improved crop yields [37]. This explains the observation that, on average, integrations of organic N inputs with mineral fertilizers gave higher net benefits than the recommended rate of fertilizer (60 kg ·N ·ha^{-1}). Considering that farmers may not be able to afford the high cost of purchasing mineral fertilizers, the use of organic inputs or their integrations with mineral fertilizers may form a major supplement or complement to replenishing nutrient deficiencies and ensuring high crop yields.

The integration of mineral fertilizers with organic in- puts or sole application of organic inputs has been regarded as a more profitable alternative in low input systems, countering the large cost of fertilizers [23]. This study confirmed that sole application of organic and/or their integrations with mineral fertilizers can be an alternative to the limited use of fertilizers.

However, this strategy may be limited by the high proportions of organic inputs required, labor required for cutting, carrying and incorporating in sole application of organics [37]. Farmers are therefore encouraged to establish their own "biomass banks" within their farms so as to reduce the transport cost and labor cost that may be incurred in search for the biomass.

CONCLUSION AND RECOMMENDATIONS

The results presented herein show that the application of organic and mineral N sources improved the general soil fertility parameters with Ca, Mg and K increasing in all treatments. Organics and/or mineral N soil amendments produce better maize yields to those obtained where the recommended mineral N sources are used alone. However, the organic materials may not be avail- able in large amounts that are required for sole application. The third objective of this study was to evaluate the economic profitability of the various soil nutrient replenishment inputs. Treatments with sole calliandra and sole tithonia had the highest BCR, followed closely by manure treatment. More so, integration of organic and mineral N sources resulted to higher net benefit and BCR than the application of the recommended rate of mineral fertilizers. Farmers are therefore are encouraged to adopt the combination of organic and mineral fertilizers as they resulted in high maize grain yields and improved soil chemical properties.

ACKNOWLEDGEMENTS

The authors are grateful and greatly indebted to RUFORUM for providing financial support for the field experimentation and soil analysis. The assistance in the soil and plants analysis by the KARI-Muguga laboratory staff is highly acknowledged.

REFERENCES

1. Jaetzold, R., Schimdt, H., Hornetz, B. and Shisanya, C. (2006) Farm management handbook of Kenya. Natural conditions and farm information Vol II/C. East Kenya. Ministry of Agriculture, Nairobi.
2. Smaling, E. (1993) Soil nutrient depletion in sub-Saha- ran Africa. In: Van Reuler, H. and Prins, W., Eds., The Role of Plant Nutrients for Sustainable Food Crop Production in Sub-Saharan Africa, VKP, Leidschendam.
3. Mutegi, J.K., Mugendi, D.N., Verchot, L.V. and Kungu, J.B. (2007) Impacts of vegetative contour hedges on soil inorganic-N cycling and erosional losses in Arable Steep- lands of the Central Highlands of Kenya. In: Bationo, A., Waswa, B.S., Kihara, J., Kimetu, J., Eds. Advances in Integrated Soil Fertility Management in Sub-Saharan Africa: Challenges and Opportunities. Springer, Dordrecht, 679-689.
4. Gitari, J.N., Kanampiu, F.K. and Matiri, F.M. (1996) Maize yield gap analysis for mid altitude areas of Eastern and Central Kenya regions. Proceedings of the 5th KARI Scientific Conference, Nairobi, October 1996, 215-225.
5. Hassan, R.M., Murithi, F.M. and Kamau, G. (1998) Determinants of fertilizer use and the gap between farmers maize yield and the potential yields in Kenya. CAB International, Wallingford, 137-161.
6. Shepherd, K.D. and Soule, M.J. (1998) Soil fertility management in Western Kenya: Dynamic simulation of productivity, profitability and sustainability

at different re- source endowment levels. Agriculture, Ecosystem and Environment, 71, 131-146. doi:10.1016/S0167-8809(98)00136-4

7. Niang, A., De Wolf, J., Hansen, T., Nyasimi, M., Rommelse, R. and Mwendwa, K. (1998) Soil fertility replenishment and recapitalization project in Western Kenya. Progress report, Maseno.
8. Adiel, R.K. (2004) Assessment of on-farm adoption potential of nutrient management strategies in Chuka Division, Meru South District, Kenya. Msc Thesis, Kenyatta University, Nairobi.
9. Mugendi, D.N., Nair, P.K.R., Mugwe, J.N., O' Neill, M.K. and Woomer, P.L. (1999) Calliandra and Leucaena alley cropped with maize Part I. Soil fertility changes and maize production in the sub humid highlands of Kenya. Agroforestry Systems, 46, 39-50. doi:10.1023/A:1006288301044
10. Donovan, G. and Casey (1998) Soil fertility management in sub-Saharan Africa. Technical Paper, World Bank, Washington DC, 408.
11. Palm, C.A., Myers, R.J. and Nandwa, S.M. (1997) Or- ganic-inorganic nutrient interaction in soil fertility replenishment. In: Buresh, R.J., Sanchez, P.A. and Calhoun F., Eds., Replenishing Soil Fertility in Africa, Soil Science Society of America, Madison, 193-218.
12. Kimani, S.K. and Lekasi, J.K. (2004) Managing manures throughout their production cycle enhances their usefulness as fertilizers: A review. In: Batiano, A., Ed., Managing Nutrient Cycles to Sustain Soil Fertility in Sub-Saharan Africa, Academy of Science Publishers, Nairobi.
13. Kenya Agricultural Research Institute (1994) Fertilizer use recommendations. Vol. 1-22. Fertilizer use recommendation project. KARI, Nairobi.
14. Vanlauwe, B., Diels, J., Aihou, K., Iwuafor, E.N., Lyasse, O., Sanginga, N. and Merckx, R. (2002) Direct interactions between N fertilizer and organic matter: Evidence from trials with 15N-labeled fertilizer. In: Vanlauwe, B., Diels, J., Sanginga, N. and Merckx, R. Eds., Integrated Plant Nutrient Management in Sub-Saharan Africa: From Concepts to Practice, CAB International, Wallingford.
15. Myers, R.J.K., Palm, C.A., Cuevas, E., Gunatilleke, I.U. N. and Brossard, M. (1994) The synchronization of nu-trient mineralization and plant nutrient demand. In: Woomer P.L. and Swift, M.J. Eds., The Biological Management of Soil Fertility, John Wiley and Sons, Chichester.
16. Jama, B.A., Swinkles, R.A. and Buresh, R.J. (1997) Agronomic and economic evaluation of organic and inorganic sources of Phosphorus in Westerbn Kenya. Agro-forestry Journal, 89, 597-604.
17. Chesson, A. (1997) Plants degradation by ruminants: Parallels with litter decomposition in soils. In: Cadisch, G. and Giller, K. Eds., Driven by Nature: Plant Litter Quality and Decomposition, CAB International, Walling-ford, 47-66.
18. Palm, C.A. and Rowland, A.P. (1997) Chemical characterization of plant quality for decomposition. In: Cadisch, G. and Giller, K.E. Eds., Driven by Nature: Plant Litter Quality and Decomposition, CAB International,

Wallingford.

19. Ayuke, F.O., Rao, M.R., Swift, M.J. and Opondo-Mbai, M.L. (2004) Effect of organic and inorganic nutrient sources in soil mineral N and maize yields in Western Kenya. In: Bationo, A. Ed., Managing Nutrient Cycles to Sustain Soil Fertility in Sub-Saharan Africa, CIAT, Cali.

20. Wallace, J.S. (1996) The water balance of mixed tree- crop systems. In: Ong, C.K. and Huxley, P. Eds., Tree Crops Interactions, a Physiological Approach, CAB International, Wallingford, 73-158.

21. Mutuo, P.K., Mukalama, J.P. and Agunda, J. (2000) On- farm testing of organic and inorganic phosphorus sources on maize in Western Kenya. TSBF Report, Cali, 22.

22. Mugendi, D. N., Mucheru-Muna, M., Mugwe, J., Kung'u, J.B. and Batiano, A. (2007) Improving food production using "best bet" soil fertility technologies in the central highlands of Kenya. In: Batiano, A., Waswa, B., Kihara, J. and Kimetu, J., Eds., Advances in Integrated Soil Fertility Management in Sub-Saharan Africa, Challenges and Opportunities, Springer, Berlin. doi:10.1007/978-1-4020-5760-1_31

23. Mucheru, M., Mugendi D.N., Kung'u, J.B., Mugwe, J. and Bationo, A. (2007) Effects of organic and mineral fertilizer inputs on maize yield and soil chemical proper-ties in a maize cropping system in Meru South District, Kenya. Agroforestry Systems, 69, 189-197. doi:10.1007/s10457-006-9027-4

24. Kapkiyai, J.J., Karanja, N.K., Woomer, P.L and Qureshi, J.N. (1998) Soil organic carbon fractions in a long term experiment and the potential for their use as a diagnostic assays in highland farming's of central Kenya. African Crop Science Journal, 6, 19-28.

25. Vanlauwe, B., Diels, J., Aihou, K., Iwuafor, E.N., Lyasse, O., Sanginga, N. and Merckx, R. (2002) Direct interactions between N fertilizer and organic matter: Evidence from trials with 15N-labeled fertilizer. In: Vanlauwe, B., Diels, J., Sanginga, N. and Merckx, R., Eds., Integrated plant Nutrient Management in Sub-Saharan Africa: From Concepts to Practice, CAB International, Wallingford.

26. Asghar, M. and Kanehiro, Y. (1980) Effects of sugarcane trash and pineapple residue on soil pH, redox potential, extractable Al, Fe and Mn. Tropical Agriculture, 57, 245- 258.

27. Hue, N.V. (1992) Correcting soil acidity of a highly weathered ultisol with chicken manure and sewage sludge. Communications in Soil Science and Plant Analysis, 23, 241-264. doi:10.1080/00103629209368586

Chapter 8

BIOMASS DIGESTION TO PRODUCE ORGANIC FERTILIZERS: A CASE-STUDY ON DIGESTED LIVESTOCK MANURE

Alessandra Trinchera, Carlos Mario Rivera, Andrea Marcucci and Elvira Rea

Agricultural Research Council - Research Centre for Plant-Soil System (CRA-RPS)., Rome, Italy

INTRODUCTION

Biogas production by anaerobic digestion of organic wastes coming from agricultural practices is one of most promising approach to generate renewable energy, giving as end-product a digested organic biomass with specific characteristics useful for soil fertilization. This last aspect represents an opportunity in relation to the need to close the nutrient cycles within the agricultural and natural ecosystems, particularly in specific systems underwent to a constant resources depletion, as those of Mediterranean area, where the C-sink loss represents one of the main causes of desertification [1], [2]. The composting process was yet identified as one of the promising answers to the need of soil organic matter conservation, such as the addition to the soil of different organic materials of different origins [3], but the anaerobic digestion could represent an effective further step able to guarantee the recycle of nutrients, coupled with an environmental-friendly energy production [4].

Particularly, anaerobic digestion of livestock manures allows us to achieve several purposes: i) renewable energy generation; ii) reduction of nitrate leaching in livestock exploitations, iii) production of an organic biomass as by-product employable as organic fertilizer [5]. Actually, digestates coming from livestock manure give biomasses characterized by biologically stable organic matter and relevant nitrogen content; these traits suppose that these biomasses may be usefully utilized as N-fertilizers and soil organic amendments in agriculture, but also as a component of growing media in pot horticultural cultivation.

It should be remarked that production of greenhouse horticultural crops requires the use of growing media of high quality, with specific physical-chemical characteristics. Being the peat, organic component traditionally used in substrates formulation, a non-renewable resource (its extraction involving many environmental issues), it becomes ever more urgent the need to individuate alternative organic materials with the same functioning, such as composts or products coming from biogestion processes [6]. Nevertheless, the use of biodigestate as amendment, still now not allowed in Italy, should provide the declaration of stability parameters for organic matter, since the utilization of matrices not properly stabilized could lead to the risk of high fermentescibility of organic components and, thus, consequent phytotoxicity phenomena [6],[7]. The stabilization of organic matter actually involves the mineralization of the most labile organic fraction, with the following decrease of C/N ratio; this means physical, chemical and biological changes of the starting material and, thus, the decrease of porosity, increase of pH, CEC, bulk density and salinity, due to a concentration of organic compounds which, generally, are characterized by lower molecular weight respect to the starting ones, more resistant to microbial degradation [8].

The amendment properties of composts and biodigestates could be assessed by different analytical methods, such as the is electrophoretic techniques (IEF) [9],[10],[11]. Results obtained after IEF characterization of the extractable fraction in alkaline environment of dried vine vinasse (an anaerobically digested solid residues, constituted by exhausted stems, skins and grape seeds, obtained after distillation of the "grappa", the Italian "acquavite"), showed an increase of the extractable organic components in alkaline environment, with a higher content in less acidic organic fraction, probably due to a "concentration effect" of the more complex organic components not, or only partially, degraded during the anaerobic digestion process.

Other works demonstrated also that the same dried vine vinasse, applied together with other mineral components in growing media, was able to increase nutrient availability [12] and express a sort of biostimulant activity on plant roots [13]. Study performed by optical microscopy demonstrated that digested vine vinasse in combination with clinoptilolite addition, promoted maize roots development, by increasing mucigel production by root tip and thus favoring the following solubilization and uptake of nutrients by plant from the added organic biomass (Figure 1).

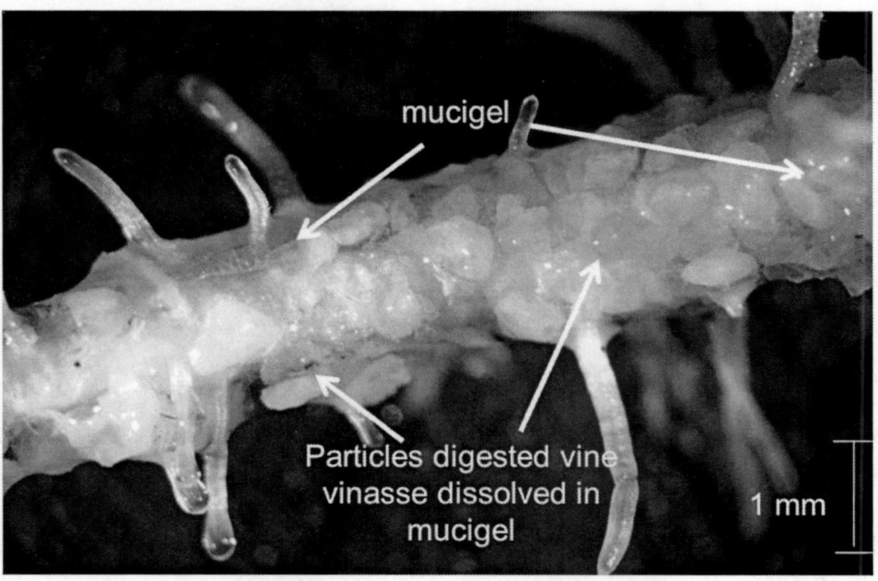

Figure 1. Root of Zea mais L., treated with micronized clinoptilolite and digested vine vinasse.

In relation to the N-fertilizer attitude of a biodigestate, it has a particular relevance in the case of the biodigestion of animal manure. The anaerobic process allows to maintain constant the total N amount of the original material, even if the organic N is mainly transformed into ammonia, the mineral form most easily available to the crops. The separation between the liquid and the solid fractions after biodigestion allows to recover the ammonium-N in the fluid fraction and the residual organic matter in the solid fraction, so to emphasize the different characteristics of the two fractions: the first one, usable as a typical N-fertilizers, able to furnish nutrient supply to plant; the second one as organic amendment, able to supply organic matter to soil and then improve its chemical, physical and biological characteristics. A proper composting process applied to the solid fraction of this digestate could lead to a further improvement of the biomass, by promoting its biochemical stability and giving those amendment properties yet described, which constitute the adding value of the final product.

What is relevant is that the risk of nitrate leaching in water represents the main limitation to the direct application of not pre-treated livestock manure to soil: effectively, its amendment properties, linked to soil organic matter addition, often go in conflict with the Council Directive 91/676/EEC on the "Water protection from nitrates" [14], especially in vulnerable zones, such as those of Italian northern regions. In this sense, the biodigestion process represents a great opportunity to utilize livestock manure digestates as N fertilizer, potentially allowing to overcome the limit of 170 Kg N ha^{-1} $year^{-1}$ superimposed by EU Nitrates Directive, since the high N-efficiency coefficient of these digestates, similar to that of soluble mineral fertilizers [15]. The solid fraction, containing organic matter already partially stabilized, could also permit its application during the winter season, provided the obtained amendment has a constant composition and a fraction of slow release-N, eventually increased by a following composting process [16],[10],[17]. On the other hand, the anaerobic conditions ensure the formation of high amount of ammonium during the organic matter degradation process, without incurring in the subsequent oxidation into nitrate [18],[4]. Being the ammonium the N-form more rapidly assimilated by the crops, this could be a further element in favor of the utilization of biodigestates as components of growing media, effectively conjugating the physical amendment properties with those chemical, connected with fertilization.

In 2007, the Italian ministry of agriculture financed a Research Project on the "Anaerobic digestion of livestock manure and EU Nitrates Directive - Effect due to the anaerobic digestion on N availability in livestock manure for overcoming the limit of 170 Kg N ha^{-1} $year^{-1}$ superimposed by imposto EU Nitrates Directive". The main aim of the reported study was to verify the N-fertilizer properties of a digestate coming from a swine livestock manure, taking into account the possibility to utilize this processed agricultural waste of animal origin for limiting the risk of environmental pollution. Hereafter, results related to the effect of the digested and not-digested solid fraction of this bovine livestock manure applied as N-organic fertilizers on lettuce growth in a greenhouse experiment are reported.

MATERIALS AND METHODS

Soils Characterization

Two different soils of Northern Italy were chosen in a specific area located in Pedemont Region (Vercelli, Italy) characterized by cold and humid climate, in relation to their recognized vulnerability for nitrate leaching. The two soils, Tetto Frati (A) and Poirino (B), sampled at 0-30 cm, were characterized for main chemical, physical and biochemical parameters: pH, CEC (meq 100 g^{-1}), total organic C % [19] and total N (mg Kg^{-1}) [20], texture (sand %, loam %, clay %), bulk density (g cm^{-3}), basal respiration $[(mg_{C\text{-}CO2} \times Kg_{soil}^{-1}) \times day^{-1}]$, C microbial biomass content (mg kg^{-1}) [21], metabolic quotient qCO_2 $[(mg_{C\text{-}CO2} \times mg\ C_{biom}^{-1} \times h) \times 10^{-3}]$, mineralization quotient qM, [22] and C microbial biomass/C organic ratio (%) [23].

BIOMASSES CHARACTERIZATION

The organic biomasses, not-processed and processed, were supplied by the anaerobic digestion plant of the "Research Centre for Animal Production - Foundation Centre for Studies and Research" (C.R.P.A. - F.C.S.R. S.p.A., Reggio Emilia, Italy). Their main chemical characteristics of solid fractions of swine livestock manure are below reported:

- pH: not-digested = 7.0 digested = 8.0
- N content (as received): not-digested = 0.47% digested = 0.39%
- N content (dry matter) not-digested = 2.9% digested = 3.0%
- C organic (dry matter): not-digested = 43,5% digested = 35,9%

In order to evaluate the organic matter stability of the two biomasses, both the digested and not-digested solid fractions of swine manure were previously analyzed by isoelectrofocusing (IEF) technique. Organic matter extraction was carried out on 2 g of each biomass with 100 mL of a solution $NaOH/Na_4P_2O_7$ 0.1N for 48 hours at 65°C. After centrifugation and filtration, the extracts were stored at 4°C under nitrogen atmosphere. Ten millilitres of $NaOH/Na_4P_2O_7$ extracts were dialysed in 6,000-8,000 Dalton membranes, lyophilised and then electrofocused in a pH range 3.5-8.0, on a polyacrylamide (acrylamide/bis-acrylamide: 37.5/1) slab gel, using 1 mL of a mixture of carrier ampholytes (Pharmacia Biotech) constituted by: 25 units of Ampholine pH 3.5-5.0; 10 units of Ampholine pH 5.0-7.0; 5 units of Ampholine pH 6.0-8.0. A prerun (2h 30'; 1200V; 21mA; 8W; 1°C) was performed and the pH gradient formed in the slab gel was checked by a specific surface electrode. The electrophoretic run (2h; 1200V; 21mA; 8W; 1°C) was carried out loading the same C amount of water-resolubilised extracts (1 mg $C \times 50\ L^{-1} \times sample^{-1}$). The bands obtained were stained with an aqueous solution of Basic Blue 3 (30%) for 18 h and then scanned by an Ultrascan-XL Densitometer (Amersham – Pharmacia) [10].

Pot Trial on Lettuce (Experimental Plan, Plant Biometric Survey And Elemental Analysis)

In a 300 m^2 greenhouse, lettuce (Lactuca sativa L., var. Romana) were transplanted into 2 L and 16 cm diameters pots, containing the A and B soils; growing density was 16 pots m^2. The experiment was performed from April to June, 2009, at temperatures range of 15-28°C.

Treatments consisted in a factorial combination of two increasing N doses (200 and 400 $kg_N \times ha^{-1}$), applied as solid fraction of digested swine livestock manure, not-digested solid fraction of swine livestock manure and granular urea [$CO(NH_2)_2$], taken as conventional mineral fertilization; not-fertilized soils were also considered as control treatments. Even if the N recommended dose for lettuce growth is about 90-100 $kg_N \times ha^{-1}$, the choice of such high

N supply was made on the basis of the need to overcome the limit of 170 Kg_N ha^{-1} by substituting these organic biomasses rich in N to the mineral fertilization, without incurring in undesired effects on plant and environment: the dose of 400 $kg_N \times ha^{-1}$ was just applied for evaluating its potential phyto-toxicity on lettuce in relation to the different fertilization treatments. The corresponding fertilizers'doses per pot were: 128.6 g and 257.2 g for ND, 153.2 g and 306.3 g for D and 1.4 g and 2.6 g for urea.

Treatments were arranged in a randomized complete-block design with six replicates. Drip irrigation was managed in relation to plant water-demand, as reported in Figure 2.

Figure 2. Example of pot cultivation of lettuce in greenhouse; drip irrigation was used for guaranteeing daily water supply to the plants.

Lettuce plants were harvested 70 days from sown; biomass dry weight (g), dry matter (%), leaf area (cm^2) and leaf number were determined for each plants. In order to evaluate the effect of alternative fertilization treatments on micro and micronutrient uptake by lettuce, N, P, K, Mg, Cu, B, Fe, Mn leaf contents, plant material was incinerated at 400°C for 24 hours; ashes were then redissolved in HCl 0.1N and the supernatant filtrated to obtain a limpid solution; the nutrient content was determined by simultaneous plasma emission spectrophotometer (ICP-OES) on obtained solution and calculated in relation to dry matter.

N USE EFFICIENCY

After analysis of N leaf tissue content (%) by Kijeldhal method, N-use efficiency (NUE %) was calculated, as the percentage of the N uptaken by the lettuce plant respect to N supplied by the fertilizer.

In order to study the long-term effect of the alternative fertilization approaches, the soil residual N at the end of experiment was obtained after Kjeldhal digestion and titrimetric determination [20]. Then, the available N-NO_3 and exchangeable N-NH_4 in the soils were determined after extraction of 4 g of each soil in 40 mL of KCl 0.2 N solution and subsequent colorimetric analysis of the supernatant by Automatic Analyzer Technicon II.

STATISTICAL ANALYSIS

Plant biometric and soil N data were evaluated by ANOVA to verify the statistical differences of the tested parameters in relation to the different fertilization treatments. Elemental data were analysed using vector analysis, which allows the simultaneous evaluation of plant dry weight and nutrients content in an integrated graphic format [24],[25]. Elemental data in relation to the different treatments in soils A and B were normalized with respect to urea al 200 $kg_N ha^{-1}$, taken as reference treatment.

RESULTS AND DISCUSSION

Soils Characterization

In Table 1, the chemical-physical and biochemical parameters of the two considered A and B soils are reported.

The comparison between soils characteristics showed that A was a typical sandy-loam soil, with a lower organic C, lower C microbial biomass content, lower C microbial biomass/C organic ratio and higher mineralization quotient respect to B, which was a loamy soil, with higher organic C, higher C microbial biomass content, higher C microbial biomass/C organic ratio and lower mineralization quotient (Figure 1). On the basis of these results, the two soils were defined as a low biological fertility soil (Soil A) and a medium biological fertility soil (Soil B) [[23],[26].

The choice of these two different soils was performed in order to evaluate the effect played by the different soil characteristics on the behavior of the digested and not-digested materials utilized in the experiment: it is well known that in all biologically mediated processes, such as the mineralization/immobilization of N coming from organic fertilizers, the microbial biomass has a key-role in addressing the degradation of organic substrate and making available N to the plant [27],[28],[29].

On the basis of what above discussed, the results were elaborated by considering the two soils as two independent experimental units, where the same experimental design was applied.

Table 1. Main chemical, physical and biochemical parameters of A and B soils.

Parameter	Soil A	Soil B
Sand (%)	48,4	15,8
Loam (%)	43,1	75,6
Clay (%)	8,5	8,6
pH	8,2	6,1
C organic (g kg^{-1})	12	17
N total (mg kg^{-1})	0,83	0,81
CEC (meq $100g^{-1}$)	8,2	12,5
Bulk density (g cm^{-3})	1,34	1,45
Basal respiration ($mg_{C\text{-}CO2}$ Kg_{soil}^{-1}) day^{-1}	7,6	5,7
C microbial biomass (mg kg^{-1})	172	281
C microbial biomass/C organic (%)	1,43	1,65
qCO_2 ($mg_{C\text{-}CO2}$ mg_{Cbiom}^{-1} h) 10^{-3}	1,85	0,85
qM (%)	4,3	3

BIOMASSES CHARACTERIZATION

In Figure 3, isoelectric focusing of the extracted organic matter from not-digested and digested solid fraction of swine manure utilized in the experimental activity is reported.

As reported in literature [30],[31], the more humified is the organic matter, the less acidic and higher in molecular weight are the organic compounds which compose it. This information, which is valid for soil humic matter but also for humic-like compounds in biomasses, allows to use IEF technique in order to separate organic matter extracted from different materials. Taking into account that, generally, organic matter focused in a pH range between 3.0 and 5.5, obtained results indicated that the organic matter from digested solid fraction was better stabilized respect to the not-digested one: in fact, the increasing peaks at pH >4.7 indicated a good stabilization of extracted organic matter from digested material, that means increased amendment properties of this material [16].

Effectively, the increase of the peaks'area in the pH region higher than 4.5 indicates firstly, that during the biodigestion changes in chemical composition of the organic material took place and, secondly, that these transformations led to the constitution of a final biomass made by less acidic compounds, higher average molecular weight molecules, that means more chemically stable material [10].

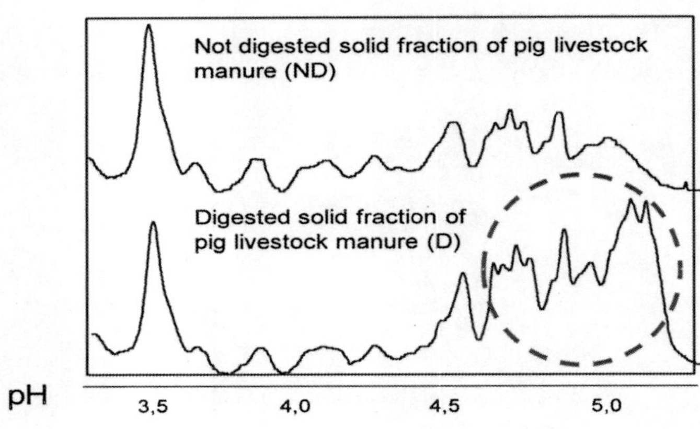

Figure 3. Isoelectrofocusing of extracted organic matter in alkaline environment from not-digested and digested solid fraction of pig livestock manure.

POT TRIAL ON LETTUCE (PLANT BIOMETRIC SURVEY AND ELEMENTAL ANALYSIS)

In Figure 4, lettuce plants differently fertilized in the two soils at the end of the experiment (60 days) are shown.

The effect of the different treatments and soils is evident when comparing the lettuce growth. The phytotoxic effect of urea at highest dose was dramatically shown in Soil A, while at the same urea dose, in soil B the lettuce was able to increase its development although the excess of N mineral supply: this result attested clearly the role of soil in influencing the actual availability of nutrients, and in particular of N, for plant uptake. It was promising the good performances of both the not-digested and the digested solid fraction of livestock manure at the highest N dose, particularly in Soil A: even if the 400 $kg_N \times ha^{-1}$ supplied by urea gave the worst result on lettuce, apparently the excess of N added with the organic biomasses did not determine any decrease of lettuce growth, but on the contrary, a very good development of lettuce foliage (Figure 5). This aspect is a positive point in order to propose the increase of the limit rate of 170 $kg_N \times ha^{-1}$ $year^{-1}$for digestate application to soil, especially because these results were obtained in a soil with a sandy texture, low organic C content and, consequently, particularly vulnerable for nitrates.

(ND = not-digested solid fraction of livestock manure; D = digested solid fraction of livestock manure; 200 = 200 $kg_N \times ha^{-1}$; 400 = 400 $kg_N \times ha^{-1}$).

Figure 4. Lettuce grown in relation to the different fertilization treatments in soil A and B.

(400 = 400 $kg_N \times ha^{-1}$).

FIgure 5. Example of lettuce leaves development in relation to the different fertilization treatments in Soil A.

The quantitative results related to biomass production and quality of plants are reported in Table 2.

Table 2. Lettuce dry weight, dry matter, total leaf area and number of leaves obtained after fertilization treatments in A and B soils (average value; different letters means significant differences at P-level<0.05).

Soil A	Dry weight (g plant^{-1})	Dry matter (%)	Total leaf area (cm^{-2})	Number of leaves
Not fertilized	1,4 b	5,2 b	450 b	15 a
Urea 200 Kg ha^{-1}	2,9 d	4,8 b	1450 d	24 c
Urea 400 Kg ha^{-1}	0,4 a	5,4 c	180 a	13 a
Not digested 200 Kg ha^{-1}	1,2 b	4,4 a	730 c	17 b
Not digested 400 Kg ha^{-1}	2,5 c	7,3 d	750 c	18 b
Digested 200 Kg ha^{-1}	1,9 c	5,0 b	760 c	17 b
Digested 400 Kg ha^{-1}	2,4 c	5,0 b	850 c	18 b
Soil B	**Dry weight (g plant^{-1})**	**Dry matter (%)**	**Total leaf area (cm^{-2})**	**Number of leaves**
Not fertilized	0,7 a	1,9 a	900 b	18 ab
Urea 200 Kg ha^{-1}	3,5 c	3,7 c	2000 c	23 c
Urea 400 Kg ha^{-1}	0,8 a	5,2 cd	1300 bc	22 c
Not digested 200 Kg ha^{-1}	1,7 b	3,7 c	850 b	18 ab
Not digested 400 Kg ha^{-1}	1,6 b	3,2 b	1250 bc	20 b
Digested 200 Kg ha^{-1}	0,7 a	2,1 a	700 a	17 a
Digested 400 Kg ha^{-1}	1,6 b	3,5 c	1000 b	20 b

Firstly, again a strong "soil effect" was recorded in relation to plant growth parameters for all the treatments, due to the different chemical-physical characteristics and biological fertility of the two soils. Not fertilized plants showed a limited vegetative development while, as expected, lowest urea dose (200 $kg_N \times ha^{-1}$) gave the best plant growth (6.7 g $plant^{-1}$), as confirmed by recorded parameters as shoot dry weight, percentage of dry matter, leaf area and leaf number, especially in B Soil. On the contrary, urea at 400 $kg_N \times ha^{-1}$ dramatically depressed plant growth in Soil A (0.8 g $plant^{-1}$), due to evident toxicity phenomena.

Digested and not digested biomasses gave best results when applied at the higher dose respect to the lowest one; actually, in treatments with both the biomasses at 400 $kg_N \times ha^{-1}$, plant parameters were closer to those obtained with urea at 200 $kg_N \times ha^{-1}$. It is relevant that in Soil A the application of both the digested and the not-digested solid fractions of livestock manure at 400 $kg_N \times ha^{-1}$ gave weight parameters higher than those observed at 200 $kg_N \times ha^{-1}$. Otherwise, in Soil B, only fertilization with digested solid fraction of livestock manure at 400 $kg_N \times ha^{-1}$ gave an increase of all the tested parameters respect to the 200 $kg_N \times ha^{-1}$ dose, while the not digested biomass did not show any differences among the N rates.

No toxicity phenomena were detected also at the highest doses of added biomasses and this is an important and positive result in the scope of utilizing these materials in substitution of mineral fertilizer also with high doses of N supply without collateral effect on plant.

For better clarify the effect of the alternative fertilization treatments on lettuce plant parameters in relation of the two soils, radar graphs are reported in Figure 6.

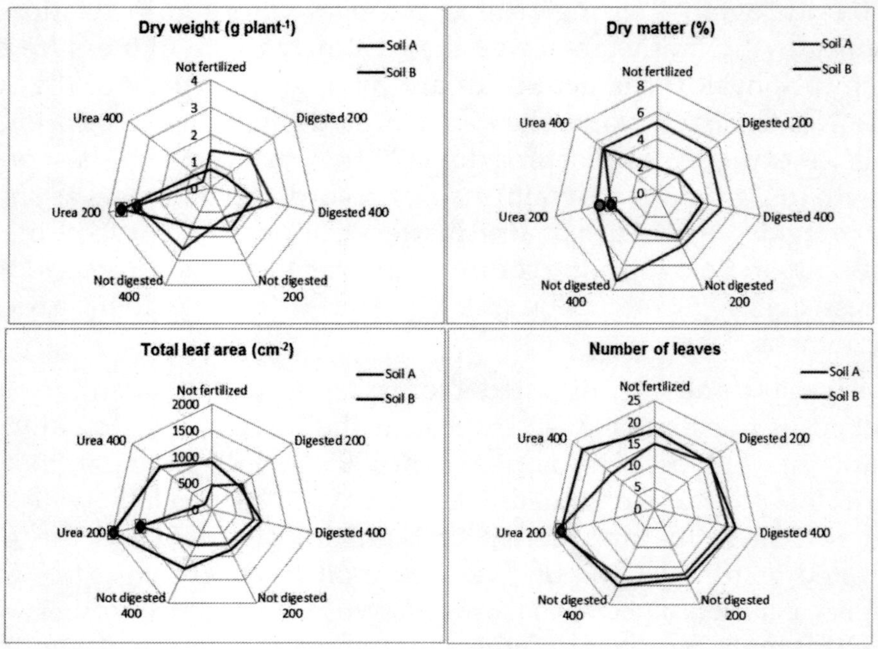

(200 = 200 $kg_N \times ha^{-1}$; 400 = 400 $kg_N \times ha^{-1}$).

Figure 6. Radar graphs of plant parameters related to the fertilization treatments in soil A and B.

Lettuce number of leaves was little affected by soil characteristics, while dry weight, dry matter and total leaf area were evidently influenced by both the soil and the fertilization. On general terms, in the Soil A the percentages of dry matter of lettuce were higher respect to the corresponding values recorded in Soil B; on the contrary, the total leaf areas were lower in Soil A than in Soil B. The water uptake seems to have had a great importance in the two considered systems: probably, it was strongly affected by physical characteristics of the two soils. In Soil A, with about 50% of sand content, the water was less available to plant because of its lower water retention capacity respect to Soil B, so to determine the tendency to reduce the lettuce total leaf area and increase leaves dry matter. On the contrary, in the loamy Soil B, the higher water availability determined a decrease in percentages of lettuce dry matter and the increase of leaf areas, so to give an indication of the dependence of lettuce quality mainly from soil characteristics rather than the fertilization treatments. Anyway, taking into account 200 $kg_N \times ha^{-1}$ of urea as the reference dose for lettuce production, it is relevant that the not-digested solid fraction of livestock manure at 400 $kg_N \times ha^{-1}$ gave the highest value of lettuce dry matter among all the treatments.

For evaluating macro (N, P, K, Mg) and micronutrients (Cu, B, Fe, Mn) use efficiency, biomass dry weight and elemental concentrations were plotted, including curved content isoclines [32],[33]. Each point on the bidimensional plot represent a vectors, taken as control equal to 100% both the concentration and the related dry weight obtained after addition of 200 $kg_N \times ha^{-1}$ urea (intersection point) in graphs (Figure 7 and Figure 8).

Plant tissue composition was significantly affected by the different treatments, depending on the added materials and rate. In relation to both the macro and the micronutrients, it should be remarked that all results were shifted to the limiting vector space (left side of the plot, under 100% lettuc

REFERENCE

1. C. M Monreal, H Dinel, M Schnitzer, D. S Gamble, V. O Biederbeck, 1997Impact of carbon sequestration on functional indicatores of soil quality as influenced by management in sustainable agriculture. ISBN: CRC Press LLC. 30435457

2. S Canali, A Trinchera, F Intrigliolo, L Pompili, L Nisini, S Mocali, B Torrisi, 2004Effect of long term addition of composts and poultry manure on soil quality of citrus orchards in Southern Italy. Biol. Fert. Soils. 40206210

3. K Paustian, R Conant, S Ogle, E Paul, 2002Environmental and management drivers of soil organic C stock changes. Proc. of the OECD Expert Meeting on soil organic C indicators for agricultural land. Ottawa, Canada.

4. A. M Mcneill, J Eriksen, L Bergstrom, K. A Smith, H Marstorp, H Kirchmann, I Nilsson, 2005Nitrogen and sulphur management: challenges fororganic sources in temperate agricultural systems. Soil Use and Man. 218293

5. P Weiland, C Rieger, T Ehrmann, 2003Evaluation of the newest biogas plants in Germany with respect to renewable energy production, greenhouse gas reduction and nutrient management. Future of Biogas in Europe II, Esbjerg 24October 2003.

6. E Rea, B De Lucia, A Ventrelli, F Pierandrei, S Rinaldi, A Salerno, L Vecchietti, V Ventrelli, 2007Substrati alternativi a base di compost per l'allevamento in contenitore di specie ornamentali mediterranee. Giornata tematica CIEC "Substrati di coltivazione: sviluppi qualitativi, tecnici, legislativi e commerciali", Milan (Italy), January 1819

7. P Sequi, E Rea, A Trinchera, 2007Aspetti legislativi per la normazione dei substrati di

coltivazione. Giornata tematica CIEC "Substrati di coltivazione: sviluppi qualitativi, tecnici, legislativi e commerciali", Milan (Italy), January 1819

8. F Lemaire, 1995Physical, chemical and biological properties of growing medium. Acta Horticol. 396273284

9. A Trinchera, A Benedetti, M Antonelli, S Salvatori, L Nisini, 2007Organic matter characterisation of amended soils under crop rotation in Mediterranean area. Geoph. Res. Abs. 9: 7635.

10. A Trinchera, F Tittarelli, F Intrigliolo, 2007Study of organic matter evolution in citrus compost by isoelectrofocusing technique. Comp. Sci. Util. 152101110

11. A Trinchera, C. M Rivera, A Marcucci, S Rinaldi, P Sequi, E Rea, 2010Assessing agronomical performances of digested livestock manure to front nitrate leaching in soil. Proceedings of the 18th European Biomass Conference and Exhibition: From Research to Industry and Markets, Lyon, May 3-7 2010. 22232227

12. A Trinchera, M Allegra, G Roccuzzo, E Rea, S Rinaldi, P Sequi, F Intrigliolo, 2011Organo-mineral fertilizers from glass-matrix and organic biomasses. A new way to release nutrients. J. Sci. Food Agr. 911323862393

13. A Trinchera, C. M Rivera, S Rinaldi, A Salerno, E Rea, P Sequi, 2010Granular size effect of clinoptilolite on maize seedlings growth. Open Agr. J. 42330

14. Council Directive 91/676/EEC concerning the protection of waters against pollution caused by nitrates from agricultural sources for the period 2004-2007.

15. I. K Thomsen, J. F Hansen, V Kjellerup, B. T Christensen, 1997Effects of cropping system and rates of nitrogen in animal slurry and mineral fertilizer on nitrate leaching from a sandy loam. Soil Use and Man. 95358

16. F Tittarelli, A Trinchera, F Intrigliolo, A Benedetti, 2002Evaluation of organic matter stability during the composting process of agroindustrial wastes. Microbiology of composting: H Insam, N Riddeck, S Klammer (Eds.) 397406

17. Y. A Abdullahi, J. C Akunna, N. A White, P. D Hallett, R Wheatley, 2008Investigating the effects of anaerobic and aerobic post-treatment on quality and stability of organic fraction of municipal solid waste as soil amendment. Biores. Technol. 86318636

18. RodrõÂguez Andara ALomas Esteban JM (1999Kinetic study of the anaerobic digestion of the solid fraction of piggery slurries. Biom. and Bioen. 17435443

19. U Springer, J Klee, 1954Prüfung der Leistungsfähigkeit von einigen wichtigeren Verfahren zur Bestimmung des Kohlemstoffs mittels Chromschwefelsäure sowie Vorschlag einer neuen Schnellmethode. Z. Pflanzenernähr. Dang. Bodenk. 64: 1.

20. Italian Official Methods for Soil AnalysisOfficial Italian Gazette, 21/10/199, n. 248.

21. E. D Vance, P. C Brookes, D. S Jenkinson, 1987An extraction method for measuring microbial biomass C. Soil Biol. Biochem. 19703707

22. Y Dommergues, 1960La notion de coefficient de minéralisation du carbone dans les sols. Agron. Trop. XV(1):54-60.

23. A Trinchera, F Pinzari, A Benedetti, P Sequi, 1999Use of biochemical indexes and changes in organic matter dynamics in a Mediterranean environment: a comparison between soils under arable and set-aside management. Org. Geochem. 30453459

24. D. L Haase, R Rose, 1995Vector analysis and its use for interpreting plant nutrient

shifts in response to silvicultural treatments. Forest Sci. 415666

25. K. I Swift, R. P Brockley, 1994Evaluating the nutrient status and fertilization response potential of planted spruce in the interior of British Columbia. Can. J. For. Res. 24594602
26. A Trinchera, F Pinzari, A Benedetti, 2001Should we be able to define soil quality before "restoring" it? Use of soil quality indicators in Mediterranean ecosystems. Minerva Biotech. 131318
27. A Benedetti, F Baroccio, A Trinchera, 2009Characterization and efficiency of slow release fertilizers. Proceedings of 16th Nitrogen Workshop, June,28th-July, 1st 2009, Turin (Italy). Grignani C., Acutis M., Zavattaro L., Bechini L., Bertora C., Gallina P.M., Sacco D. (Eds.). 393394
28. A Benedetti, F Baroccio, A Trinchera, S Mocali, 2009Potential mineralization of soil organic nitrogen in aerobic and anaerobic conditions. Proceedings of 16th Nitrogen Workshop, June,28th-July, 1st 2009, Turin (Italy). Grignani C., Acutis M., Zavattaro L., Bechini L., Bertora C., Gallina P.M., Sacco D. (Eds.). 2324
29. A Trinchera, P Nardi, A Benedetti, 2010Influence of biological fertility of soil on methylenurea biodegradation. Proceedings of 18th Symposium of the International Scientific Centre of Fertilizers, 8-12 November 2009, Rome (Italy). 506510
30. M Govi, C Ciavatta, C Gessa, 1994Evaluation of the stability of the organic matter in slurries, sludges and composts using humification parameters and isoelectric focusing. Humic Substances in the Global Environment and Implications on Human Health. Senesi S and Miano TM (Eds). Elsevier Science. 13111316
31. M Govi, C Ciavatta, D Montecchio, P Sequi, 1995Evolution of organic matter during stabilization of sewage sludge. Agr. Med., 125107114
32. D. W Valentine, H. L Allen, 1990Foliar responses to fertilization identify nutrient limitation in loblolly pine. Can. J. For. Res. 20144151
33. C. F Scagel, 2003Growth and nutrient use of ericaceous plants grown in media amended with sphagnum moss peat or coir dust. Soil Management, fertilization and irrigation. HortSci. 3814654

Chapter 9

IMPACT OF ORGANIC AND INORGANIC FERTILIZERS APPLICATION ON THE PHYTOCHEMICAL AND ANTIOXIDANT ACTIVITY OF KACIP FATIMAH (LABISIA PUMILA BENTH)

Mohd Hafiz Ibrahim [1,*], Hawa Z. E. Jaafar [2], Ehsan Karimi [2] and Ali Ghasemzadeh [2]

[1] Department of Biology, Faculty of Science, University Putra Malaysia, Serdang 43400, Selangor, Malaysia

[2] Department of Crop Science, Faculty of Agriculture, University Putra Malaysia, Serdang 43400, Selangor, Malaysia; E-Mails: hawazej@gmail.com (H.Z.E.J.); ehsan_b_karimi@yahoo.com (E.K.); upmali@yahoo.com (A.G.)

ABSTRACT

A study was conducted to compare secondary metabolites and antioxidant activity of Labisia pumila Benth (Kacip Fatimah) in response to two sources of fertilizer [i.e., organic (chicken dung; 10% N:10% P_2O_5:10% K_2O) and inorganic fertilizer (NPK green; 15% N, 15% P_2O5, 15% K_2O)] under different N rates of 0, 90, 180 and 270 kg N/ha. The experiment was arranged in a randomized complete block design replicated three times. At the end of 15 weeks, it was observed that the application of organic fertilizer enhanced the production of total phenolics, flavonoids, ascorbic acid, saponin and gluthathione content in L. pumila, compared to the use of inorganic fertilizer. The nitrate content was also reduced under organic fertilization. The application of nitrogen at 90 kg N/ha improved the production of secondary metabolites in Labisia pumila. Higher rates in excess of 90 kg N/ha reduced the level of secondary metabolites and antioxidant activity of this herb. The DPPH and FRAP activity was also highest at 90 kg N/ha. The results indicated that the use of chicken dung can enhance the production of secondary metabolites and improve antioxidant activity of this herb.

INTRODUCTION

Medicinal plants are used to cure many ailments that are either non-curable or seldom cured through modern systems of medicine. Approximately 80% of the World population depends on medicinal plants for their health and healing [1]. Societal motivations to use herbs are increasing due to concern about the side effects of synthetic drugs. Many botanicals and some dietary supplements are good sources of antioxidants and anti-inflammatory compounds [2]. Labisia pumila Benth., commonly called 'Kacip Fatimah', of the family Myrsinaceae is an important medicinal herb which has been used in traditional Malay medicine since time immemorial. The roots, leaves or the whole plant are boiled in water [3,4] and the water taken orally to facilitate labor, shrink the uterus and improve menstrual irregularity and for post-partum medicine [5,6]. Its usage for women's health may be related to its phytoestrogen effects and content of chemicals with similar structure to estrogen [7]. Labisia pumila extract contains flavanoids, phenolics and various bioactive volatile compounds [8]. Chemical analysis of its roots revealed three novel metabolites, demethylbelamcandaquinone B, fatimahol and dexylo-primulanin, together with 21 known compounds, including epoxyoleanane glycosides, alkenated phenolics, cerebroside, glycerogalactolipids, and lipids [9].

For healthy growth and optimal yield, nutrients must be available to plants in correct quantity, proportion and in a usable form at the right time. To fulfill these requirements, chemical fertilizers and/or organic manures are needed. Fertilization has been reported to have an influence on the phyto-nutritional quality of crops. Inorganic fertilizer is said to reduce the antioxidant levels, while organic fertilizer has been proven to enhance antioxidant content in plants [10]. Applying fertilizers, particularly in the inorganic form, in excess of plant requirements can increase the chances of fertilizer loss and environmental pollution. Organic manures, apart from improving physical and biological properties of soil, help in improving the efficiency of chemical fertilizers [11]. Organic manures such as farmyard and poultry manure are known to improve the physical, chemical and biological conditions of soil and ensure sustainable soil health [12].

In the past, agricultural production was focused on maximizing the quantity of crop produced for commercial markets. Hence, compound fertilizer has been used as a common agricultural practice.

However, recently health conscious consumers are interested in optimizing the nutritional composition with minimal chemical residues on foods produced through environmentally friendly agricultural practices [13]. Substituting chemicals with organic fertilizers is one of the common principles in this production system. Inorganic fertilizers have had significant effects on World crop production and are essential components of today's agriculture. Estimates show that agricultural production is raised by 50% as a result of chemical fertilizers and 60% of the population owes its nutritional survival to nitrogen (N) fertilizers [14]. However, of the total applied N, less than 50% is recovered in the soil–plant system, while the remainder is lost to the environment [15]. Plants can take up nitrogen (N) either as inorganic ions (NH_4^+ or NO_3^-), or as organic N [16–18]. Gao et al. [18] observed that application of high NH_4^+ and low NO3−levels resulted in improved fruit quality. There is a scarcity of data on the effect of the form of N on the production of secondary metabolites in plants [19]. According to the C/N balance hypothesis, when N is readily available, plants will primarily make compounds with high N content (e.g., proteins for growth). When N availability is limited, metabolism changes more towards carbon-containing compounds such as starch cellulose, and non-N-containing secondary metabolites such as phenolics and terpenoids [20]. The relative differences in the release of nutrients from various fertilizers could lead to different C/N ratios in plants and this in turn could lead to a difference in the production of secondary metabolites [17]. Recently, Hakkinen and Torronen [21] compared the phenolic content in three cultivars of strawberries grown organically and conventionally. They reported that only one cultivar grown under organic conditions showed higher levels of phenolics than its inorganically grown counterparts. Asami et al. [22] reported significantly higher total phenolics in marionberries grown with organic fertilizer as compared with inorganic fertilizer. Weibel et al. [23] indicated that the phenol (mainly flavonols) content of organically grown apple cultivars was 19% higher than in apples grown using inorganic fertilizers. These results suggest that the usage of organic fertilizers can enhance the production of plant secondary metabolites.

The fertilizer impact on vegetative growth is well documented. However, the effect of fertilizer rates and sources on phytochemical quality in L. pumila Benth is lacking. There have been numerous studies that have investigated the effects of various environment factors on plant primary and secondary metabolites of this herb [24–34], but only limited studies covering the response of secondary metabolites and antioxidant activities under different fertilizer sources and rates. Hence, the objective of this study was to examine the effect of fertilizer source (chicken dung and NPK green) and fertilizer rates (0, 90, 180 and 270 kg N/ha) on the production of secondary metabolites (total phenolics, flavonoid, phenolics, ascorbic acid, saponin) and antioxidant activity (DPPH and FRAP) of L .pumila var alata seedlings. The relationships between these parameters were also investigated.

RESULTS AND DISCUSSION

Total Phenolics and Flavonoids

Total phenolics and flavonoids were influenced by fertilizer source and fertilization rates ($p \leq 0.05$). It was observed that that application of organic fertilizer increased the production of total phenolics and flavonoids contents in L. pumila Benth. (Table 1). Total phenolics and flavonoids were enhanced 12% and 22%, respectively, in the organic fertilizer treatment compared to inorganic fertilization. It was apparent that optimum fertilization ocurred at 90 kg N/ha, where it was observed that total phenolics and flavonoids had the highest values (total phenolics = 1.32 mg/g gallic acid dry weight, and total flavonoids = 0.81 mg/g rutin dry weight). Increased fertilization rates above 90 kg N/ha to 180 and 270 kg N/ha resulted in a decrease in the total phenolics and flavonoid contents. Ferry et al. [35] and Ranelletti et al. [36] have also shown that thesephenolics and flavonoids had anticancer activities, and that they were able to inhibit cancer cell growth. Gallic acid and rutin were reported to have high scavenging activity and act as treatments for diabetes, albuminaria, psoriasis and external haemorrhoids [37,38]. Some studies have reported that gallic acid and rutin play an important role in the prevention of cancer [39]. The present results suggest that the application of organic fertilizers can enhance L. pumila antimicrobial and anticancer activities. The results indicated that total phenolics and flavonoid had a positive significant correlation with soluble sugar (total phenolics = r^2= 0.912; total flavonoid r^2 = 0.923; $p \leq 0.05$) (Table 2). This indicated that under low fertilization rates there will be an accumulation of soluble sugar that might enhance the total phenolics and flavonoids in *Labisia pumila*. Previous studies done on Labisia pumila have indicated that carbohydrate and production of total phenolics and flavonoid was positively correlated and this was also observed in the present study [24–34].

Table 1. Total phenolics, flavonoids, vitamin c and saponin concentrations in leaf tissue of L. pumila in response to fertilizer rates and source.

Treatments	Total phenolics (mg gallic acid/g dry weight	Total flavonoids (mg rutin/g dry weight)	Vitamin C (mg/g fresh weight)	Saponin (mg diosgenin/g dry weight)
Fertilizer sources				
Organic (Chicken dung)	1.10a	0.76a	0.061b	38.16a
Inorganic (NPK green)	0.98b	0.62b	0.078a	32.17b
Fertilizer rates (kg N/ha)				
0	1.22b	0.81b	0.072b	47.21b
90	1.32a	0.98a	0.089a	58.14a
180	1.02c	0.72c	0.060c	32.18c
270	0.87d	0.41d	0.047d	20.17d
Source × Rate	ns	ns	ns	ns

ns: non significant at $p \leq 0.05$. Means followed by the same letters are not significantly different by DNMRT test ($p \leq 0.05$).

Table 2. The correlationship among the parameters recorded in the study (total phenolics, total flavonoids, vitamin c, saponin, soluble sugar, nitrate, gluthathione, DPPH and FRAP).

Parameters	1	2	3	4	5	6	7	8	9
1. T. phenolics	1.000								
2. T. flavonoids	0.899 *	1.000							
3. Vitamin C	0.768 *	0.777 *	1.000						
4. Saponin	0.912 *	0.813 *	0.776 *	1.000					
5. Soluble sugar	0.912 *	0.923 *	0.954 *	0.914 *	1.000				
6. Nitrate	0.212	0.123	0.231	0.321	0.216	1.000			
7.Gluthathione	0.914 **	0.924 *	0.889 *	0.824 *	0.778 *	0.123	1.000		
8. DPPH	0.912 *	0.915 *	0.911 *	0.976 *	0.918 *	0.126	0.908 *	1.000	
9. FRAP	0.899 *	0.917 *	0.931 *	0.876 *	0.854 *	0.421	0.912 *	0.987 *	1.000

Ascorbic Acid and Saponin Content

The same trend was also observed for ascorbic acid and saponin contents in Labisia pumila. It was observed that ascorbic acid and saponin content was maximized at 90 kg N/ha. The increase in fertilization rates (0 > 270 kg N/ha) reduced the production of ascorbic acid and saponin contents in this plant. The reduction in ascorbic acid and saponin content under high fertilization was also reported in other studies [40–42]. Saponin content was highest under organic fertilization that recorded

38.16 mg/g fresh weight, and lowest with inorganic fertilization which only registered 32.17 mg/g fresh weight. However, ascorbic acid was found to be highest with inorganic fertilization. The usage of inorganic fertilization increased the ascorbic acid content by 27% compared to the usage of organic fertilizer. It was observed that organic fertilization enhanced the production of total phenolics, flavonoids and saponin contents in L. pumila (Table 2). The increase in production of secondary metabolites under organic fertilization in the present study might be due to high in micronutrients content in the organic fertilizer used. Plants grown under organic agricultural conditions are reported to have higher micronutrient content in more cases than conventionally grown plants [23,24].

Considering the fact that some of chemical reactions in cells involve minor elements, either directly or indirectly, this could explain why organic-fertilized plants exhibited higher production of secondary metabolites [43].

Soluble Sugar and Nitrate

Soluble sugar content was influenced by fertilizer type and rates ($p \leq 0.05$). It was observed that soluble sugar content was higher under organic fertilization compared to inorganic fertilization. At the end of 15 weeks the soluble sugar content in the organic fertilizer treatment was 97.2 mg/g dry weight compared to 92.10 mg/g dry weight recorded with the inorganic fertilization. Less soluble sugar was produced in 0 kg N/ha (98.10 mg/g dry weight), 180 kg N/ha (71.25 mg/g dry weight) and 270 kg N/ha (62.31 mg/g dry weight) treatments compared to fertilization at 90 kg N/ha, which produced the highest soluble sugar accumulation (102.70 mg/g dry weight). This phenomenon was also observed by other researchers [44–46]. The accumulation of soluble sugar under low nitrogen fertilization might be due to reduction in sink size of the plant when nitrogen is limited, hence reducing the translocation of carbohydrate to the other plant parts [47]. The current results showed that high application rates of fertilizer increased the nitrate content in leaf sap samples (Table 3). It was also observed that the nitrate content was statistically lower in the treatment with organic fertilizer compared to inorganic fertilizer. The nitrate content with inorganic fertilizer was 13% more than with organic fertilizer.

Nitrate has been attributed to negative effects to human health. Toxicity of nitrate to human can be manifested by headaches, syncope, vertigo and discoloration that manifest in fingers or lips [48]. The results indicate that the usage of organic fertilizer can minimize the nitrate content in plants. Organic fertilizer contains nitrogen bound to organic material that is slowly released [49]. Inorganic fertilizer which is absorbed rapidly into the plant usually have higher nitrosamines which have been associated with chronic diseases such as leukemia and gastrointestinal cancer [50]. The nitrate levels that acceptable to human are 10–100 ppm in drinking water. Intake more than 100 ppm would induce the toxic effects [51]. In the current result, the nitrate value for organic and non-organic was higher than acceptable recommended levels (>100 ppm). However, most vegetables and fruits contained nitrate levels from

20–200 ppm [52]. Usually, the processing of food (e.g., chopping, grinding and heating) would reduce the nitrate content and thus reduce the negative effects [53]. The current result implies that the application of organic fertilizer to L. pumila can reduce the nitrate content and plays an important role in the production of healthier herbal products [54].

Table 3. Total soluble sugar, nitrate and gluthathione concentration in leaf tissue of L. pumila in response to fertilizer rates and source

Treatments	Soluble sugar (mg sucrose/g dry weight)	Nitrate (ppm)	Gluthathione (nmol/g dry weight)
Fertilizer sources			
Organic (Chicken dung)	97.21a	157.21b	616.71b
Inorganic (NPK green)	90.02b	178.79a	632.16a
Fertilizer rates (kg N/ha)			
0	98.17b	112.31b	777.21b
90	102.72a	137.21a	831.28a
180	71.25c	198.21c	621.31c
270	62.31d	211.31d	522.18d
Source x Rate	ns	ns	ns

ns: non significant at $p \leq 0.05$. Means followed by the same letters are not significantly different by DNMRT test ($p \leq 0.05$).

Gluthathione Content

The gluthathione levels were influenced by fertilizer source and rates ($p \leq 0.05$). The production of gluthathione fertilized at 270 kg N/ha was significantly lower than other fertilization rates. At the end of 12 weeks of measurement, gluthathione for 0, 90 and 180 kg N/ha was 771.21, 831.28 and 621.31 nmol compared to only 522.18 nmol for 270 kg N/ha. The gluthathione levels at 0, 90 and 180 kg n/ha were 47, 59 and 19% higher than with fertilization at 270 kg N/ha. GSH is a tri-peptide composed of cysteine, glutamic acid and glycine and is the most abundant non-protein thiol in the cells. Its active group is the thiol (–SH) of cysteine. GSH is maintained in the reduced state. The GSH plays an imperative role in the stabilization of many antioxidant enzymes [55]. Additionally, as an antioxidant scavenger it serves as a substrate for dehydroascorbate (DHAsA) reductase and is also directly reactive with free radicals including the hydroxyl radical to prevent the inactivation of enzymes by oxidation of an essential thiol group [56]. In the present study, we found that reduced N fertilization increased GSH. High GSH is necessary for several physiological functions, including activation and inactivation of redox-dependent enzyme systems and regeneration of the cellular antioxidant ascorbic acid under oxidative conditions [57,58]. Usually, the increase in GSH with reduced N fertilization is associated with an increase in antioxidant properties [59]. In the current study, it was shown that GSH had a strong positive relationship with total phenolics, flavonoids, ascorbic acid and saponin content (Table 3). The results showed that the increase in antioxidative properties of L. pumila under low nitrogen fertilization might be due to an increase in production of total phenolics, flavonoids, ascorbic acid, saponin, and that GSH activity can increase the antioxidant capacity of this plant under this condition [60,61].

DPPH and FRAP

The effects on DPPH are attributed to fertilizer source and rates of nitrogen levels ($p \leq 0.05$; Table 4). At 380 µg/mL, the DPPH antioxidant activity recorded the highest value (63.18%) at 90 kg N/ha followed by the 0 kg N/ha (57.12%), 180 kg N/ha (47.22%), and the least in the 270 kg N/ha treatment (38.99%). The DPPH values were found to be highest under organic fertilization (47.28%) and lowest under inorganic fertilization (40.21%). The results imply that the usage of organic fertilizer can enhance the radical scavenging activity in L. pumila and high fertilizer rates could significantly reduce the DPPH radical scavenging activity of the medicinal plant. It must be noted that the DPPH assay principally measures the activity of water-soluble antioxidants [62]. The principle of this method is that in the presence of a molecule consisting of a stable free radical (DPPH), an antioxidant with the ability to donate a hydrogen atom will quench the stable free radical, a process which is associated with a change in the absorption and can be measured spectrophotometrically.

Results of the current work also suggest that high N supply was disadvantageous for the improvement of the antioxidant activity of water-soluble antioxidants in L. pumila. Besides phenolics and flavonoid compounds, other water-soluble antioxidants in the extracts such as vitamin C could also exert an additive effect on DPPH radical scavenging activity. Studies have shown that a combination of phenolics and ascorbic acid produced a synergistic effect on DPPH radical scavenging activity [63].

Table 4. DPPH and FRAP scavenging assay in leaf tissue of L. pumila in response to fertilizer rates and source.

Treatments	DPPH scavenging assay (%)	FRAP scavenging assay (μm fe(II)/g dry weight)
Fertilizer sources		
Organic (Chicken dung)	47.28a	600.24a
Inorganic (NPK green)	40.21b	528.17b
Fertilizer rates (kg N/ha)		
0	57.12b	701.24b
90	63.18a	814.21a
180	47.22c	621.71c
270	41.10d	511.78d
Source × Rate	ns	ns

ns: non significant at $p \leq 0.05$. Means followed by the same letters are not significantly different by the DNMRT test ($p \leq 0.05$).

The Ferric Reducing Antioxidant Potential (FRAP) assay measures the total antioxidant power of biological fluids [64]. Total antioxidant power was assessed by the reduction of Fe^{3+} to Fe^{2+}, which occurred rapidly with all reductants with half of the reaction reduction potentials above that of Fe^{3+}/Fe^{2+}. Therefore, the values express the corresponding concentration of electron-donating antioxidants. The FRAP was influenced by fertilizer source and rates ($p \leq 0.01$), and followed the same trend as with DPPH, where the reducing ability was highest under organic fertilization with the activity maximized at 90 kg N/ha, and was the lowest activity with 270 kg N/ha. The present results indicate that fertilization with low N enabled high abilities to reduce ferric Ions. The antioxidant potential of L. pumila was estimated by the ability to reduce 2,4,6-tripyridyl-s-triazine (TPTZ)-Fe(III) complex to TPTZ-Fe(II). The ferric reducing ability (FRAP assay) has been widely used in the evaluation of the antioxidant component of dietary polyphenols [65]. The antioxidant activity is found to be linearly proportionate to the phenolics and flavonoids content [66–69]. Yen et al. [70] reported that the ferric reducing power of bioactive compounds was associated with antioxidant activity. Glenn et al. [71] reported a strong positive relationship between total phenolics, flavonoids compounds and antioxidant activity. A similar trend was observed with the current study where total phenolics and flavonoids displayed significantly positive relationships with FRAP activity ($r^2 = 0.899$ and $r^2 = 0.917$; $p \leq 0.05$; Table 2). Furthermore, DPPH and FRAP had a significant positive relationship with GSH and ascorbic acid, and thus justifies that high DPPH and FRAP activity in L. pumila extracts under organic fertilization at low rates might be due to high accumulation of total phenolics, total flavonoids, gluthatione and ascorbic acid in the plant.

The positive impact of chicken dung (organic fertilizer) applied at the lower rates may be due to increased metabolic activity under these conditions [72]. The usage of organic fertilizer improves soil properties by increasing soil physical, chemical and biological properties, but the usage also prevents soil erosion [73]. The increase in secondary metabolites and antioxidant activity with organic fertilizer might be due to the availability of various other major and minor elements, whereas the inorganic fertilizer used in the current study only supplied the three major elements [Nitrogen (N;15%), Potassium (P_2O_5; 15%) and Phosphorous (K_2O; 15%)]. The current study indicated that supplying L. pumila with chicken dung (organic fertilizer) improves the secondary metabolites production and antioxidant activity of this plant especially under low N (<90 kg N/ha).

EXPERIMENTAL

Experimental Location, Plant Materials and Treatments

The experiment was carried out in a glasshouse complex at Field 2, Faculty of Agriculture Glasshouse Complex, Universiti Putra Malaysia (longitude 101° 44′ N and latitude 2° 58′ S, 68 m above sea level) with a mean atmospheric pressure of 1.013 kPa. The seedlings were planted in a soilless medium containing coco-peat, burnt paddy husk and well composted chicken manure in a 5:5:1 (v/v) ratio in 25 cm diameter polyethylene bags. The media had a pH value of 6.00. Day and night temperatures were maintained at 27–30 °C and 18–21 °C, respectively, with a relative humidity of between 50 to 60%. All the seedlings were irrigated using overhead mist irrigation, given four times a day or when necessary. Each irrigation session lasted for 7 min. Each treatment consisted of seven seedlings, and there were a total of 252 seedlings used in the experiment. Three-month old *L. pumila* var alata seedlings were left for a month in the nursery to acclimatize until they were ready for the treatments. When the seedlings had reached 4 months of age they were fertilized with one of two fertilizer sources [i.e., chicken dung based BIO-ORGANIC® organic fertilizer (10% N: 10% P_2O_5: 10% K2O) and inorganic fertilizer NPK green (15% N, 15% P_2O_5, 15% K_2O)] and were evaluated at different N rates of 0, 90, 180 and 270 kg N/ha. This factorial experiment was arranged in a randomized complete block (RCBD) design with three replications. All the plants were harvested after 15 weeks of treatment.

Determination of Total Phenolics and Flavonoids

The methods used for extraction and quantification of total phenolics and flavonoids contents followed that described in Ibrahim et al. [26]. A fixed amount of ground tissue samples (0.1 g) was extracted with 80% ethanol (10 mL) on an orbital shaker for 120 min at 50 °C. The mixture was subsequently filtered (Whatman™ No.1), and the filtrate was used for the quantification of total phenolics and total flavonoids. Folin–Ciocalteu reagent (diluted 10-fold) was used to determine total phenolics content of the leaf samples. Two hundred μL of the sample extract was mixed with

Folin–Ciocalteau reagent (1.5 mL) and allowed to stand at 22 °C for 5 min before adding $NaNO_3$ solution (1.5 mL, 60 g L^{-1}). After two hours at 22 °C, absorbance was measured at 725 nm. The results were expressed as mg g^{-1} gallic acid equivalent (mg GAE g^{-1} dry sample). For total flavonoids determination, samples (1 mL) were mixed with $NaNO_3$ (0.3 mL) in a test tube covered with aluminium foil, and left for 5 min. Then 10%

$AlCl_3$ (0.3 mL) was added followed by addition of 1 M NaOH (2 mL). The absorbance was measured at 510 nm using a spectrophotometer with rutin as a standard (results expressed as mg/g rutin dry sample).

Total Saponin Determination

Total saponin content was determined according to Makkar and Becker [74] based on the vanillin-sulfuric acid colorimetric reaction. The results were expressed as mg diosgenin equivalent per gram dry matter of plant material.

Ascorbic Acid Determination

The ascorbic acid content was measured using a modified method of Davis and Masten [75]. The fresh leaf samples (1 g) were extracted in 1% of phosphate-citrate buffer, pH 3.5 using a chilled pestle and mortar. The homogenate was filtered. The filtrate was added to 1.7 mM 2,6-dichloroindophenol (2,6-DCPIP, 1 mL) in a 3 mL cuvette. The absorbance at 520 nm was read within 10 min of mixing the reagents. The extraction buffer was used as a blank. L-Ascorbic acid was used as a standard. Ascorbic acid was recorded as mg/g L-ascorbic acid in fresh leaves.

Soluble Sugar Determination

Soluble sugar was measured spectrophotometrically using the method of Ibrahim et al. [25]. Samples (0.5 g) were placed in 15 mL conical tubes, and distilled water added to make up the volume to 10 mL. The mixture was then vortexed and incubated for 10 min. Anthrone reagent was prepared using anthrone (Sigma Aldrich, St. Louis, MO, USA, 0.1 g) that was dissolved in 95% sulphuric acid (Fisher Scientific, Los Angeles, CA, USA 50 mL). Sucrose was used as a standard stock solution to prepare a standard curve for the quantification of sucrose in the sample. The mixture of ground dry sample and distilled water was centrifuged at a speed of 3400 rpm for 10 min and then filtered to get the supernatant. A sample (4 mL) was mixed with anthrone reagent (8 mL) and then placed in a water-bath at 100 °C for 5 min before the sample was measured at an absorbance of 620 nm using a spectrophotometer (Model UV160U; Shimadzu

Scientific, Kyoto, Japan). The total soluble sugar in the sample was expressed as mg/g sucrose in dry sample.

Nitrate Determination

Fresh leaf samples were collected and kept in a refrigerator prior to analysis. The fresh leaves were cut to small pieces and squeezed in a stainless steel press to obtain the sap. Sap was then used to measure nitrate concentration using a Horiba® Cardy Twin Nitrate Meter (Model B-343; Horiba Scientific Inc., Trenton, NJ, USA).

Gluthathione Determination

Total glutathione were determined by reacting plant extracts (0.5 mL) with 50 mM KH2PO4/2.5 mM EDTA [28]. buffer (pH 7.5), 0.6 mM DTNB [5,5-dithiobis-2-nitrobenzoic acid] in 100 mM Tris-HCl, pH 8.0, 1 unit of glutathione reductase (GR, from spinach, EC 1.6.4.2) and 0.5 mM NADPH. GSH was quantified from the reaction mixture by mixing plant extracts (0.5 mL) with 60 mM KH2PO4/2.5 mM EDTA buffer (pH 7.5), 0.6 mM DTNB [5,5-dithiobis-2-nitrobenzoic acid] in 200 mM Tris-HCl, pH 8.0. The mixture was incubated at 30 °C for 15 min, and the rate of change in absorbance was determined at 412 nm using a light spectrophotometer (UV-3101P, Labomed Inc., Lincoln, NE, USA).

DPPH Radical Scavenging Assay

The 1,1-diphenyl-2-picryl-hydrazyl (DDPH) used was purchased from Sigma-Aldrich. The DPPH free radical test was conducted using the method of Ghasemzadeh et al. [76]. The initial absorbance of DPPH in methanol was measured at 515 nm until the absorbance remained constant. Extracts (40 μL) were added to alcohol solutions of DPPH (3 mL, 0.1 mM). The samples were first kept in a dark place at room temperature and after 30 min the absorbance was measured using a spectrophotometer (U-2001, Hitachi Instruments Inc., Tokyo, Japan) at 515 nm. The percent of inhibition was determined using the formula: Percent of inhibition (%) = [(A515 of control − A515 of sample)/A515 of control] × 100.

Reducing Ability (FRAP Assay)

The ability to reduce ferric ions was measured using a modified method of Karimi et al. [77]. An aliquot (200 μL) of the extract with appropriate dilution was added to 3 mL of FRAP reagent (10 parts of 300 mM sodium acetate buffer at pH 3.6, 1 part of 10 mM TPTZ solution and 1 part of 20 mM FeCl3 ·6H2O solution) and the reaction mixture was incubated in a water bath at 37 °C. The increase in absorbance at 593 nm was measured after 30 min. The antioxidant capacity based on the ability to reduce ferric ions of the extract was expressed as μM Fe(II)/g dry mass.

Statistical and Correlation Analysis

Data were analyzed using the analysis of variance procedure in SAS version 17. Means separation between treatments was performed using Duncan multiple range test and the standard error of differences between means was calculated with the assumption that data were normally distributed and equally replicated [78–82]. Correlation analysis were analyzed using Pearson correlation analysis to establish the relationship between all the variables.

CONCLUSION

In the present study, it was observed that the use of chicken dung (organic fertilizer) resulted in higher production of secondary metabolites and increased antioxidant activity compared to the use of NPK 15:15:15 (inorganic fertilizer). The production of total phenolics, flavonoid, ascorbic acid, saponin and antioxidant activity was highest under low nitrogen fertilization, especially at 90 kg N/ha. It was also observed that higher rates of fertilizer enhanced the production of nitrate. The nitrate content was found to be the lowest with chicken dung fertilization. The results indicated that the use of chicken dung was superior compared to the NPK fertilizer in producing high quality herbal plants (L. pumila) for pharmaceutical use.

ACKNOWLEDGEMENTS

The authors are grateful to the Ministry of Higher Education Malaysia for financing this work under the Research University Grant Scheme No. 91007.

REFERENCES

1. Aliyu, L. Effect of manure type and rate on growth, yield and yield components of pepper. J. Sustain. Agric. Environ. 2003, 5, 92–98.
2. Balasubramanian, P.; Palaniappan, S.P. Principles and practices of agronomy; Tata McGraw-Hill Publishing Co. Ltd.: New Delhi, India, 2001.
3. Zainon, A.S.; Musaadah, M.; Ismail, M.; Wan-Fadhilah, W.Z.A. Ethobotany of Medicinal Plants at Pos Lanai, Lipis, Pahang. In Interdisciplinary Approaches in Natural Products Research; Mawardi, K., Zhari, A., Suparjo, M., Chong, S., Eds; Department of Chemistry, University Putra Malaysia: Serdang, Malaysia, 1999; pp. 35–42.
4. Runi, S.P. Studies on Medicinal Plant in Sarawak. In Towards Bridging Science and Herbal Industry; Chang, Y.S., Mastura, M., Vimala. S., Zainon, A.S., Eds.; Forest Research Institute of Malaysia (FRIM): Kuala Lumpur, Malaysia, 2001; pp. 112–119.
5. Zakaria, M.; Mohammed, M.A. Traditional Malay Medicinal Plants; Fajar Bakti: Kuala Lumpur, Malaysia, 1994.
6. Burkill, I.H. Dictionary of the Economic Products of the Malay Peninsula; Crown Agents for the Colonies: London, UK, 1935.
7. Jamia, A.J.; Houghton, J.P.; Milligan, R.S.; Jantan, I. The oestrogenic and cytotoxic effects of the extracts of Labisia pumila var. alata and Labisia pumila var. pumila in vitro. Med. Sci. J. 2003, 1, 53–60.
8. Karimi, E.; Jaafar, H.Z. HPLC and GC–MS determination of bioactive compounds in microwave obtained extracts of three varieties of Labisia pumila Benth. Molecules 2011, 16, 6791–6805.
9. Ali, Z.; Khan, I.A. Alkyl phenols and saponins from the roots of Labisia pumila (Kacip Fatimah). Phytochemistry 2011, 72, 2075–2080.
10. Dumas, Y.; Dadomo, M.; Di Lucca, G.; Grolier, P. Effects of environmental factors and agricultural techniques on antioxidant content of tomatoes. J. Sci. Food Agric. 2003, 83, 369–382.
11. Abd-Alla, H.M.; Yan, F.; Schubert, S. Effects of sewage sludge application on nodulation, nitrogen fixation and plant growth of faba bean, soybean and lupin. J. Appl. Bot. 1999, 73, 69–75.
12. Logan, T.J.; Lindsay, B.J.; Goins, L.E.; Ryan, J.A. Field assessment of sludge metal bioavailability to crops: Sludge rate response. J. Environ. Qual. 1997, 26, 534–550.

13. Fabiyi, L.L.; Ogunfowora, O.O. Economics of Production and Utilization of Organic Fertilizer. In Organic Fertiliser in Nigerian Agriculture: Present and Future; Federal Ministry of Science and Technology: Lagos, Nigeria, 1992; pp. 138–144.

14. Follet, R.H.; Murphy, L.S.; Donalue, R.L. Soil-fertilizer-plant relationship. Fert. Soil Amend. 1981, 6, 478–481.

15. Barker, A.V. Organic vs. inorganic nutrition and horticultural crop quality. Hortscience 1975, 10, 50–53.

16. Gagnon, B.; Berrouard, S. Effects of several organic fertilizers on growth of greenhouse tomato transplants. Can. J. Plant Sci. 1994, 74, 167–168.

17. Montagu, K.D.; Goh, K.M. Effects of forms and rates of organic and inorganic nitrogen fertilizers on the yield and some quality indices of tomateos (Lycopersicon esculentum Miller). N. Z. J. Crop Hort. Sci. 1990, 18, 31–37.

18. Gao, Z.; Sagi, M.; Lips, H. Assimilate allocation priority as affected by nitrogen compounds in the xylem sap of tomato. Plant Physiol. Biochem. 1996, 34, 807–815.

19. Brandt, K.; Molgaard, P. Organic agriculture: does it enhance or reduce the nutritional value of plant foods? J. Sci. Food Agric. 2001, 81, 924–931.

20. Haukioja, E.; Ossipov, V.; Koricheva, J.; Honkanen, T.; Larsson, S.; Lempa, K. Biosynthetic origin of carbon-based secondary compounds: Cause of variable responses of woody plants to fertilization? Chemoecology 1998, 8, 133–139.

21. Hakkinen, S.H.; Torronen, A.R. Content of flavonols and selected phenolic acids in strawberries and Vaccinium species: Influence of cultivar, cultivation site and technique. Food Res. Int. 2000, 33, 517–524.

22. Asami, D.K.; Hong, Y.J.; Barrett, D.M.; Mitchell, A.E. Comparison of the total phenolic and ascorbic acid content of freeze-dried and air-dried marionberry, strawberry, and corn using conventional, organic, and sustainable agricultural practices. J. Agric. Food Chem. 2003, 51, 1237–1241.

23. Weibel, F.P.; Bickel, R.; Leuthold, S.; Alfoldi, T. Are organically grown apples tastier and healthier? A comparative field study using conventional and alternative methods to measure fruit quality. Acta Hort. 2000, 517, 417–426.

24. Ibrahim M.H.; Jaafar, H.Z.E. Photosynthetic capacity, photochemical efficiency and chlorophyll content of three varieties of Labisia pumila Benth exposed to open field and greenhouse growing conditions. Acta Physiol. Plant. 2011, 33, 2179–2185.

25. Ibrahim, M.H.; Jaafar, H.Z.E.; Asmah, R.; Zaharah, A.R. Involvement of Nitrogen on flavonoids, glutathione, anthocyanin, ascorbic acid and antioxidant activities of Malaysian medicinal plant Labisia pumila Blume (Kacip Fatimah). Int. J. Mol. Sci. 2012, 13, 393–408.

26. 26. Ibrahim, M.H.; Jaafar, H.Z.E.; Rahmat, A.; Abdul Rahman, Z. The relationship between phenolics and flavonoids production with total non structural carbohydrate and photosynthetic rate in Labisia pumila Benth. under high CO2 and nitrogen fertilization. Molecules 2011, 16, 162–174.

27. Ibrahim, M.H.; Jaafar, H.Z.E. The influence of carbohydrate, protein and

phenylanine ammonia lyase on up-regulation of production of secondary metabolites (total phenolics and flavonoid) in Labisia pumila (Blume) Fern-Vill (Kacip Fatimah) under high CO2 and different nitrogen levels. Molecules 2011, 16, 4172–4190.

28. Ibrahim, M.H.; Hawa, Z.E.J. Carbon dioxide fertilization enhanced antioxidant compounds in Malaysian Kacip Fatimah (Labisia pumila Blume). Molecules 2011, 16, 6068–6081.

29. Ibrahim, M.H.; Jaafar, H.Z.E. Enhancement of leaf gas exchange and primary metabolites, up-regulate the production of secondary metabolites of Labisia Pumila Blume seedlings under carbon dioxide enrichment. Molecules 2011, 16, 3761–3777.

30. Jaafar, H.Z.; Ibrahim, M.H.; Karimi, E. Phenolics and flavonoid compounds, phenylanine ammonia lyase and antioxidant activity responses to elevated CO2 in Labisia pumila (Myrisinaceae). Molecules 2012, 17, 6331–6347.

31. Ibrahim, M.H.; Jaafar, H.Z. Reduced photoinhibition under low irradiance enhanced kacip fatimah (Labisia pumila Benth) secondary metabolites, phenyl alanine lyase and antioxidant activity. Int. J. Mol. Sci. 2012, 13, 5290–5306.

32. Ibrahim, M.H.; Jaafar, H.Z.E.; Rahmat, A.; Zaharah, A.R. Effects of nitrogen fertilization on synthesis of primary and secondary metabolites in three varieties of kacip fatimah (Labisia pumila Blume). Int. J. Mol. Sci. 2011, 12, 5238–5254.

33. Jaafar, H.Z.E.; Ibrahim, M.H.; Mohamad Fakri, N.F. Impact of soil field water capacity on secondary metabolites, phenylalanine ammonia-lyase (PAL), maliondialdehyde (MDA) and photosynthetic responses of malaysian kacip fatimah (Labisia pumila Benth). Molecules 2012, 17, 7305–7322.

34. Ibrahim, M.H.; Jaafar, H.Z.E. The relationship of nitrogen and C/N on secondary metabolites and antioxidant activities in three varieties of Malaysian kacip fatimah (Labisia pumila Blume). Molecules 2011, 16, 5514–5526.

35. Ferry, D.R.; Smith, A.; Malkhandi, J. Phase I clinical trial of the flavonoid quercetin: Pharmacokinetics and evidence for in vivo tyrosine kinase inhibition. Clin. Cancer Res. 1996, 2, 659–668.

36. Ranelletti, F.O.; Maggiano, N.; Serra, F.G. Quercetin inhibits p21-ras expression in human colon cancer cell lines and in primary colorectal tumors. Int. J. Cancer 1999, 85, 438–445.

37. Pathak, S.B.; Niranjan, K.; Padh, H.; Rajani, M. TLC densitometric method for the quantification of eugenol and gallic acid in clove. Chromatographia 2004, 60, 241–244.

38. Enkhmaa, B.L. Mulberry (Morus albaL.) leaves and their major flavonol quercetin 3-(6-malonylglucoside) attenuate atherosclerotic lesion development in LDL receptor-deficient mice. J. Nutr. 2005, 135, 729–734.

39. Luo, Y.L. Inhibition of cell growth and VEGF expression in ovarian cancer cells by flavonoids. Nut. Cancer 2008, 60, 800–809.

40. Seung, K.L.; Adel, A.K. Preharvest and postharvest factors influencing vitamin C content of horticultural crops. Postharvest Biol. Technol. 2000, 20, 207–220.

41. Hassan, A.; Predrag, L.; Irina, P.; Omar, S.; Uri, C.; Arieh, B. Fertilization-induced changes in growth parameters and antioxidant activity of medicinal plants used in traditional Arab medicine. Oxf. J. 2005, 2, 549–556.

42. Toor, R.K.; Savage, G.P.; Heeb, A. Influence of different types of fertilizers on the major antioxidant components of tomatoes. J. Food Compost. Anal. 2006, 19, 20–27.

43. Bimova, P.; Pokluda, R. Impact of organic fertilizers on total antioxidant capacity of head cabbage. Hortic. Sci. (Prague) 2009, 36, 21–25.

44. Tissue, D.T.; Thomas, R.B.; Strain, B.R. Atmospheric CO2 increases growth and photosynthesis of Pinus taedea: A four year field experiment. Plant Cell Environ. 1997, 20, 1123–1134.

45. Den-Hertog, J.; Stulen, L.; Fonseca, E.; Delea, P. Modulation of carbon and nitrogen allocation in Urtica dioica and Plantago major by elevated CO2: Impact of accumulation of non-structural carbohydrates and ontogenetic drift. Physiol. Plant. 1996, 98, 77–88.

46. Poorter, H.; Berkel, V.; Baxter, R.; Den-Hertog, J.; Dijkstra, P.; Gifford, R.M.; Griffin, K.L.; Roumet, C.; Roy, J.; Wong, S.C. The effects of elevated CO2 on the chemical composition and construction costs of leaves of 27 C3 species. Plant Cell Environ. 1997, 20, 472–482.

47. Meyer, S.; Cerovic, Z.G.; Goulas, Y.; Montpied, P.; Demotes, S.; Bidel, L.P.R.; Moya, I.; Dreyer, E. Relationship between assessed polyphenols and chlorophyll contents and leaf mass per area ratio in woody plants. Plant Cell Environ. 2006, 29, 1338–1348.

48. Winchester, P.D.; Huskins, J.; Ying, J. Agrichemicals in surface water and birth defects in the United States. Acta Paediat. 2009, 98, 664–669.

49. Benbrook, C.; Zhao, X.; Yáñez, J.; Davies, N.; Andrews, P. New Evidence Confirms the Nutritional Superiority of Plant-Based Organic Foods. State of Science Review; The Organic Center: Boulder, CO, USA, 2008.

50. Rembialkowska, E. Quality of plant products from organic agriculture. J. Sci. Food Agric. 2007, 87, 2757–2762.

51. Van, D.W.; Loch, J.P.G. Nitrate in the Netherlands: A serious threat to groundwater. Aqua 1983, 2, 59–60.

52. Gangolli, S.D. Assessment: Nitrate, nitrite and N-nitroso compounds. Eur. J. Pharmacol. 1994, 292, 1–38.

53. Gatseva, P.; Dimitrov, I. Population morbidity in a community with nitrate contamination of drinking water. Folia Medica 1997, 39, 65–71.

54. Leroy, B.M.M.; Bommele, L.; Reheul, D.; Moen, M.; de Neve, S. The application of vegetable, fruit and garden waste (VFG) compost in addition to cattle slurry in a silage maize monoculture: Effects on soil fauna and yield. Eur. J. Soil Biol. 2007, 43, 91–100.

55. Dunning, S.; Ur Rehman, A.; Tiebosch, M.H.; Hannivoort, R.A.; Haijer, F.W.; Woudenberg, J.; van den Heuvel, F.A.; Buist-Homan, M.; Faber, K.N.; Moshage, H. Glutathione and antioxidant enzymes serve complementary roles in protecting activated hepatic stellate cells against hydrogen peroxide-induced cell death. Biochim. Biophys. Acta 2013, 1832, 2027–2034.

56. Dalton, D.A.; Russell, S.A.; Hanus, F.J.; Pascoe, G.A.; Evans, H.J. Enzymatic reactions of ascorbate and glutathione that prevent peroxide damage in soybean root nodules. Proc. Natl. Acad. Sci. USA 1986, 83, 3811–3813.

57. Wang, Y.S.H.; Bunce, A.J.; Maas, L.J. Elevated carbon dioxide increases contents of antioxidant compounds in field-grown strawberries. J. Agric. Food Chem. 2003, 51, 4315–4320.

58. Ziegler, D.M. Role of reversible oxidation reduction of enzyme thiol-disulfides in metabolic regulation. Annu. Rev. Biochem. 1985, 54, 305–329.

59. Lewis, N.G. Plant Phenolics. In Antioxidants in Higher Plants; Alscher, R.G., Hess, J.L., Eds.; CRC: Boca Raton, FL, USA, 1993; pp. 135–160.

60. Larson, R.A. The antioxidants of higher plants. Phytochemistry 1988, 27, 969–978.

61. Guo, R.; Yuan, G.; Wang, Q. Effects of sucrose and mannitol accumulation of health promoting components and activity of metabolic enzymes in brocolli sprout. Sci. Hort. 2011, 128, 159–165.

62. Frankel, E.N.; Huang, S.W.; Kanner, J.; German, J.B. Interfacial phenomena in the evaluation of antioxidants: Bulk oils versus emulsions. J. Agric. Food Chem. 1994, 42, 1054–1059.

63. Murakami, M.; Yamaguchi, T.; Takamura, H.; Matoba, T. Effects of ascorbic acid and tocopherol on antioxidant activity of polyphenolic compounds. Food Chem. Toxicol. 2003, 68, 1622–1625.

64. Benzie, I.F.; Strain, J.F. The ferric reducing ability of plasma (FRAP) as a measure of "antioxidant power": The FRAP assay. Anal. Biochem. 1996, 239, 70–76.

65. Luximon-Ramma, A.; Bahorun, T.; Soobrattee, A.M.; Aruoma, O.I. Antioxidant activities of phenolics, proanthocyanidin and flavonoid components in extracts of Acacia fistula. J. Agric. Food Chem. 2005, 50, 5042–5047.

66. Ibrahim, M.H.; Jaafar, H.Z.E. Abscisic Acid Induced Changes in Production of Primary and Secondary Metabolites, Photosynthetic Capacity, Antioxidant Capability, Antioxidant Enzymes and Lipoxygenase Inhibitory Activity of Orthosiphon stamineus Benth. Molecules 2013, 18, 7957-7976.

67. Ghasemzadeh, A.; Jaafar, H.Z.E.; Rahmat, A.; Wahab, P.E.M.; Halim, M.R.A. Effect of different light intensities on total phenolics and flavonoid synthesis and anti-oxidant activities in young ginger varieties (Zingiber officinale Roscoe). Int. J. Mol. Sci. 2010, 11, 3885–3897.

68. Ghasemzadeh, A.; Jaafar, H.Z.E. Effect of CO2 enrichment on synthesis of some primary and secondary metabolites in ginger (Zingiber officinale Roscoe). Int. J. Mol. Sci. 2011, 12, 1101–1114.

69. Ghasemzadeh, A.; Jaafar, H.Z.E.; Rahmat, A. Synthesis of phenolics and flavonoids in ginger (Zingiber officinale Roscoe) and their effects on photosynthesis rate. Int. J. Mol. Sci. 2010, 11, 4539–4555.

70. Yen, G.C.; Duh, P.D. Scavenging effects of methanolic extract of peanut hulls on free-radical and active oxygen species. J. Agric. Food Chem. 1994, 42, 629–632.

71. Glenn, M.I.; Thomas-Barberan, F.T.; Hess-Pirce, B.; Kader, A.A. Antioxidant capacities, phenolic compounds, carotenoids and vitamin C contents of nectarine, peach and plum cultivars from California. J. Agric. Food Chem. 2002, 50, 4976–4982.

72. Woese, K.; Lange, D.; Boess, C; Bögl, K.W. A comparison of organically and conventionally grown foods—Results of a review of the relevant literature. J. Sci. Food Agric. 1997, 74, 281–293.

73. Khalil, M.Y.; Moustafa, A.A.; Naguib, N.Y. Growth, phenolic compounds and antioxidant activity of some medicinal plants grown under organic farming condition. J. Agric. Sci. 2007, 3, 451–457.

74. Makkar, H.P.S.; Siddhuraju, S.; Siddhuraju, P.; Becker, K. Plant Secondary Metabolites; Humana Press: Totowa, NJ, USA, 2007.

75. Davies, S.H.R.; Masten, S.J. Spectrophotometric method for ascorbic acid using dichlorophenolindophenol: Elimination of the interference due to iron. Anal. Chim. Acta 1991, 248, 225–227.

76. Ghasemzadeh, A.; Jaafar, H.Z.; Rahmat, A. Elevated carbon dioxide increases contents of flavonoids and phenolic compounds, and antioxidant activities in malaysian young ginger (Zingiber officinale Roscoe.) varieties. Molecules 2010, 15, 7907–7922.

77. Karimi, E.; Ehsan, O.; Rudi, H.; Jaafar, H.Z.E. Evaluation of Crocus sativus L. Stigma phenolic and flavonoid compounds and its antioxidant activity. Molecules 2010, 15, 6244–6256.

78. Wu, H. Affecting the activity of soybean lipoxygenase-1. J. Mol. Graph. 1996, 14, 331–337.

79. Ibrahim, M.H.; Jaafar, H.Z.E.; Haniff, M.H.; Raffi, M.Y. Changes in growth and photosynthetic patterns of oil palm seedlings exposed to short term CO2 enrichment in a closed top chamber. Acta Physiol. Plant. 2010, 32, 305–313.

80. Ibrahim, M.H.; Jaafar, H.Z.E. Impact of elevated carbon dioxide on primary, secondary metabolites and antioxidant responses of Eleais guineensis Jacq. (Oil Palm) seedlings. Molecules 2012, 17, 5195–5211.

81. Ibrahim, M.H.; Jaafar, H.Z.E. Primary, secondary metabolites, H2O2, malondialdehyde and photosynthetic responses of Orthosiphon stimaneus Benth to different irradiance levels. Molecules 2012, 17, 1159–1176.

82. Ibrahim, M.H., Jaafar, H.Z.E. Relationship between extractable chlorophyll content and SPAD values in three varieties of kacip fatimah under greenhouse conditions J. Plant Nutr. 2013, 36, 1366–1372.

Chapter 10

ALTERNATE NITROGEN AMENDMENTSFOR ORGANIC FERTILIZERS

M.K.C. Sridhar[1,*], G.O. Adeoye[2], and O.O. AdeOluwa[2]

Organo-Mineral Fertilizer Research and Development Group

[1]Division of Environmental Health, College of Medicine, University of Ibadan, Ibadan, Nigeria

[2]Department of Agronomy, University of Ibadan, Ibadan, Nigeria

ABSTRACT

The use of compost or manure in agriculture as an organic source of nutrients is common in many tropical, developing countries like Nigeria. One of the drawbacks of such materials is their low nitrogen (N) content (=1% N). Farmers commonly use chemical N fertilizers such as urea, calcium ammonium nitrate (CAN), and NPK formulations to obtain better crop growth and yield. These chemical supplements may have a negative impact on the environment through nitrate leaching into water, leading to eutrophication of surface waters that can affect public health.

Gliricidia sepium, a fast-growing, tropical, perennial hedge plant was tested as a source of N in organo-mineral fertilizer formulations. Average nutrient content of Gliricidia is 3.8% N, 0.32% P1.8% K, 0.8% Ca, and 0.2% Mg. Using a sand cuture and Amaranthus caudatus as a test crop, it was shown that amending commercial composts with 30% Gliricidia prunings would benefit many small-scale farmers and control environmental pollution.

DOMAINS: plant sciences, agronomy, soil systems, eco- systems and communities, environmental chemistry, bioremediation and bioavailability, environmental technol- ogy, environmental management and policy, ecosystems management, biotechnology, agricultural biotechnology

INTRODUCTION

Before the advent of mineral fertilizers about 150 years ago, manure and composts were practically the only sources of nutri- ents for crops[1]. However, after independence and the discov- ery of crude oil in the 1960s, the use of mineral fertilizers displaced organic sources as people found them easy to handle and to apply on their farms. In 4 decades, it was recognized that several problems occurred with intensive use of inorganic fertil- izers. Some of these problems were reliability of fertilizer sup- ply, high cost, N volatilization losses, soil degradation through erosion, and frequent overapplication with consequent pollution of surface and groundwater resources. These problems have been encountered in many countries.

As a result, most countries began to put emphasis on the advantages of applying organic manure or organic fertilizers to replenish dwindling soil nutrients. Among the various organic materials in use are composted city refuse, green manure, de- composed animal wastes, and farmyard manure, which are usu- ally found in abundant quantities in most communities. In Nigeria, millions of tons of city refuse are produced annually, which has the potential to be composted and used to improve soil fertility. *Gliricidia sepium*[2,3], a common hedge plant in the humid trop- ics, was found to be useful as a source of green manure. Com- pared to other shrubs, *Gliricidia* has a high decomposition rate and N release[4]. *Gliricidia* is also productive, yielding from 2 to 15 t/ha/year of dry matter[2]. In a study carried out by Handayanto et al.[5], it was observed that the N mineralization rate of low-grade prunings from *Peltophorum dasyrrachis* was significantly improved by mixing with *G. sepium.*

Many researchers have reported the superiority of combined inorganic and organic fertilizers (organo-mineral fertilizers) over their performance separately[6,7]. However, Cooke[8] and Peverly and Gates[9] showed that organic fertilizers could indeed perform better for some crops. The use of organic matter as fertilizer in developing countries has received much less at- tention from economists because the beneficial effects are not visible immediately[10]. Araji and Stodick[11] found that the cost associated with the handling and spreading of manure ranges between 20 to 30% of the cost of commercial fertilizer. This cost differential indicates that the use of green manure can be less costly than using urea or any chemical amendment in compost. Even though the use of organo-mineral fertilizers is widely ac- cepted globally, the inorganic constituents may still pose envi- ronmental problems to some extent.

In order to minimize environmental impact, there is a need for local materials that can supplement the N and other minerals. One such alternative identified is *G. sepium,* which grows exten- sively as a wild hedge plant. This paper reports the assessment of *G. sepium* for its suitability as an amendment for organo-mineral fertilizers being prepared from market and slaughterhouse wastes

MATERIALS AND METHODS

Fertilizer Sources

- Grade B organo-mineral fertilizer (OMF): Compost was obtained from Pace Setter Organic Fertilizer Plant, developed by the authors for the Oyo state government. The compost was made from market and slaughterhouse wastes from Ibadan, in southwestern Nigeria. The unamended compost was prepared using an aerobic windrow system and was devoid of toxic and heavy metals. This is referred to as Grade B OMF.

Grade A OMF: The Grade B OMF was amended with urea and bone meal to enrich with extra N and P and is referred to as Grade A OMF.

- *Gliricidia* prunings were obtained from agricultural land by harvesting fresh leaves, drying to constant weight at a temperature of 80°C, and milling to pass through a 0.2-mm sieve.

Fertilizer Treatments

Grade A OMF, Grade B OMF, and *Gliricidia* were used in the greenhouse experiments. With these materials, seven treatments were prepared (letter-coded as shown below) and applied as follows:

- Soil alone with no treatments (control) P
- 20% dried *Gliricidia* prunings + 80% Grade B OMF A
- 30% dried *Gliricidia* prunings + 70% Grade B OMF B
- 40% dried *Gliricidia* prunings + 60% Grade B OMF C
- 50% dried *Gliricidia* prunings + 50% Grade B OMF D
- 100% dried *Gliricidia* prunings alone M
- Grade B (unamended) OMF alone N
- Grade A OMF alone O

The soil used in the experiments was obtained from the University of Ibadan campus, which from the composition may be considered as the worst of tropical eroded soil. The soil was sandy (with about 90% sand) and slightly acidic (with pH 6.6). It had very low levels of mineral nutrients and exchangeable cations. This soil was deliberately chosen for the experiment with a view of establishing the efficacy of the available nutrients from the treatments under study (Table 1). The soil was sieved to pass 2 mm to remove stones and other unwanted constituents. In each pot, 2 kg soil was mixed with various amendments. *Amaranthus caudatus* was used as the test crop.

GREENHOUSE EXPERIMENT

The greenhouse experiment was designed for evaluating the efficacy of the various treatments. Each treatment was applied at 5,

10, and 15 t/ha in 2-kg plastic pots in a randomized complete block design. Soil weight was 1.8 kg per pot. *A. caudatus* was planted twice successively, for main and residual effects. The ambient air temperature was about 28 to 30°C. *A. caudatus* was planted 1 week after the incorporation of the various treatments.

Figure 1. *G. sepium* in its natural habitat from a farm

For finding the residual effect, fresh seeds were planted immediately after the first harvest, with no additional treatments. Water was applied daily to field capacity to maintain soil moisture content. If any weeds were seen, they were removed, ground, and put back into the same pots to account for nutrient loss. The fol- lowing parameters were observed at 5 weeks after planting (WAP) and 6 WAP for residual effect: plant height, leaf number, stem girth, and whole-plant fresh and dry weight. Plant height was taken from the ground level to the dewlap of the last fully opened leaf. Stem girth was determined by passing a thread around a predetermined mark of 10 cm on the stem above the soil level. The harvested samples were oven dried to constant weight at 80°C and expressed as tons per hectare.

METHODS OF ANALYSES

Six soil core samples (passed through a 2-mm sieve) were taken randomly for physical and chemical analysis. They were air-dried and used. Particle size distribution was determined by hydrom- eter method[12] using sodium hexametaphosphate as the dispers- ing agent. The coarse-sand fraction was separated from the fine sand using a 1-mm sieve.

Soil pH was determined by suspending the soil samples in distilled water (1:1) using glass electrode (EIL pH meter)[13]. Organic carbon was analyzed by the dichromate wet-oxidation

Table 1 Physico-Chemical Properties of the Experimental Soil

Parameter	Value
pH value (in water)	6.60
Organic carbon, %	1.50
Total Kjeldahl N, g/kg	1.3
Available P, mg/kg	7.50
Exchangeable bases, cmol/kg	
K	0.13
Na	7.50
Mg	0.18
Ca	0.14
Exchangeable acidity, cmol/kg	0.20
CEC, cmol/kg	4.28
Base saturation, %	95.50
Extractable micronutrients, mg/kg	
Mn	47.98
Fe	66.90
Cu	0.85
Zn	3.55
Mechanical composition, g/kg	
Sand	910.0
Silt	74.0
Clay	20.0

method of Walkey-Black as described by Black[14], and the value of organic matter was obtained by multiplying the carbon level by 1.729. Total Kjeldahl N was determined by macro Kjeldahl method. Available P was determined by the Bray P (0.03

N NH4F + 0.025 N HCl) solution as described by Bray and

Kurtz[15] and determined colorimetrically by the molybdate blue method.

Exchangeable cations were determined by leaching with 1N neutral NH4OAc according to the method of Tisdale[16]. K was determined by flame photometer as described by Sillanpaa[17]. Micronutrients (Mn, Fe, Cu, Zn) were extracted with 0.1 N EDTA and determined using atomic absorption spec- trophotometer[18].

The results of the experiment were subjected to analysis of variance and the means compared using the Duncan Multiple Range test[19].

RESULTS AND DISCUSSION

The Grade B OMF, which is the normal compost made from urban wastes, as expected was low in N and P (Table 2). The Grade A OMF was amended with urea and bone meal to increase N and P levels, respectively. Most farmers in Nigeria prefer a formula that can supply adequate N to meet requirements of cer- tain high-N-demanding crops like maize and vegetables. P amend- ment is only needed for certain crops. In other African and Asian countries, the agricultural practice involves supplementing or- ganic manure separately with inorganic fertilizers sometime dur- ing the growing period, usually after several weeks of growth. This practice is not common in Nigeria, where traditional fal- lowing methods or green manure are used when inorganic fertil- izers are not available or cost effective.

Gliricidia has three to four times higher N content compared to normal compost (i.e., Grade B OMF). Therefore, the Grade B OMF was amended with ground dry leaves at various percent- ages from 0 to 100 along with appropriate controls (soil alone) to compare the efficacy in providing N to the test crop. Green amaranth is a widely used vegetable among the low- and middle- income traditional families. This test crop was monitored for the growth and dry-matter yield.

The highest *Amaranthus* yield at 5 WAP was 16.78 t/ha with the 30% *Gliricidia*-amended compost. Norman[20] reported a similar yield of 20 t/ha for amaranths grown on a sandy, infertile soil. If the soil is reasonably fertile, unlike the one used in this study, the yield can be substantial. Incorporation of *Gliricidia* improved the N levels of unamended compost (Table 2).

While the 100% *Gliricidia*-prunings treatment proved to be most profitable, the 30% Gliricidia amendment seems to be ideal from the practical point of view. All the Gliricidia-amended composts in deed performed better than the urea-amended composts. The optimum application rate of amended composts was 15 t/ha both in the first and second crops (Table 3). The yields observed in the second crop may be due to residual nutrient availability.

Gliricidia prunings gradually released nutrients over a period of 120 days[4]. Even the 40 and 30% Gliricidia-amended compost had significant residual effects compared to the urea-amended compost.

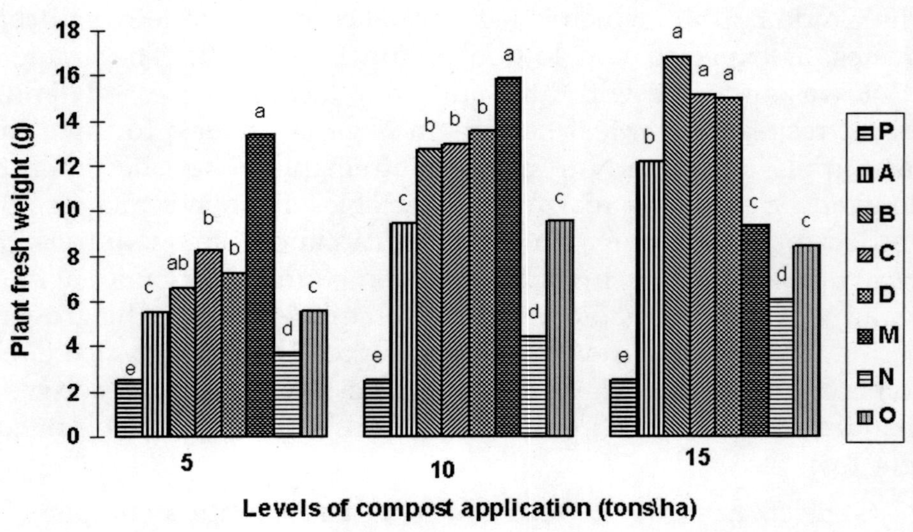

Figure 2. Effect of various fertilizers on fresh weight of Amaranthus at the end of 5 weeks after planting

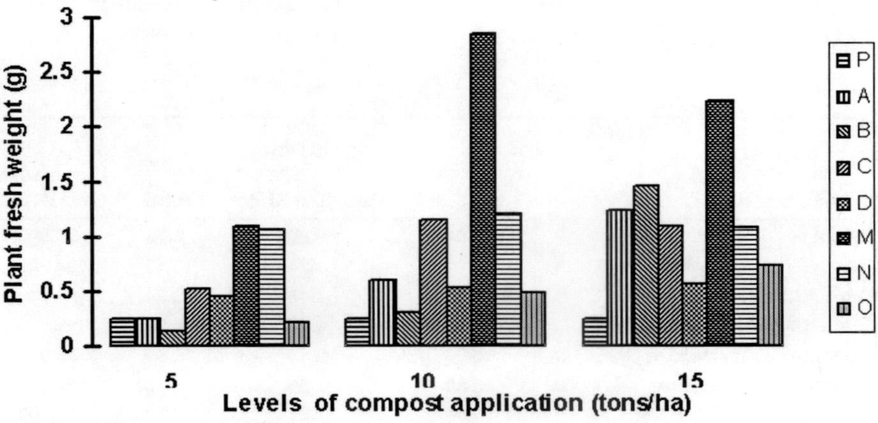

Figure 3. Residual effect of various fertilizers on fresh weight of Amaranthus at the end of 6 weeks after planting.

Table 2 Nutrient Composition of Organo-Mineral Fertilizers and G. sepium Prunings

	Material Used		
Composition	Grade A OMF	Grade B OMF	*G. sepium*
Total Kjeldahl N, %	2.58	1.46	3.78
Total phosphorus (P), %	1.10	1.03	0.32
Potassium (K), %	0.68	0.60	1.83
Calcium (Ca), %	3.62	1.19	0.80
Magnesium (Mg), %	0.18	0.11	0.20
Zinc (Zn), mg/kg	276.0	290.0	31.00
Copper (Cu), mg/kg	25.0	22.00	—
Manganese (Mn), mg/kg	32.0	19.00	—
Chromium (Cr), mg/kg	ND[a]	ND	—
Lead (Pb), mg/kg	ND	ND	—

Table 3 Estimated Income from Amaranthus Production Based on Best Level of Performance for Each Treatment

Treatments		Yield (t/ha)			Income/ha
		Main Yield	Residual Effect	Total	(NGN)[a]
Control	P	2.5	0.25	2.75	27,500
20% *Gliricidia*-amended OMF	A	12.19	1.24	13.43	134,300
30% *Gliricidia*-amended OMF	B	16.78	1.46	18.34	183,400
40% *Gliricidia*-amended OMF	C	15.16	1.09	16.25	162,500
50% *Gliricidia*-amended OMF	D	15.03	0.56	15.59	155,900
100% *Gliricidia*-pruning	M	15.88	2.85	18.73	187,300
Unamended Grade B OMF	N	6.04	1.07	7.11	71,100
Urea-amended Grade A OMF	O	9.58	0.48	10.06	100,600

[a] Currency equivalent: 1 USD = 114 NGN (Nigerian Nairas).

From an economic point of view (Table 3), the 100% *Gliricidia*-amended compost gave the best yield, and, in terms of currency, it amounted to about seven times higher than the control and almost double that of the urea-amended compost. However, it is not practicable to use the plant entirely, and the next best is 30%, which is very feasible when mixed with the composts prepared from wastes. When amended and applied at a rate of 10 t/ha, the economic returns are six times higher than the control. These results concur with Cooke[8] and Perverly and Gates[9], who reported that sole organic fertilizers performed better with some crops. Mafongoya et al.[21] studied the N-re- lease patterns from seven leguminous trees and found that the lignin + polyphenol to N ratio could be used to screen legumi- nous tree leaves for their potential to release N. In another study by Lionel et al.[22] in Haiti, *G. sepium* was found to decompose faster as compared to seven other Hedgerow species. Leaves decomposed faster than stems. These studies further confirm the value of *Gliricidia* in compost amendments.

CONCLUSIONS

The results of this study showed that *Gliricidia* prunings as an organic amendment for compost significantly improved growth and yield of *A. caudatus* compared to urea-amended compost. Cost and environmental advantages suggested that 100% *Gliricidia* prunings and their 30, 40, and 50% composition in compost could do better than urea-amended compost. Thus, the *Gliricidia* performed better in sandy soil than conventional urea used as a compost amendment. Based on results from the green- house study, the following conclusions may be drawn:

- Dry *Gliricidia* prunings are shown to be a good substitute for mineral fertilizers in enriching N levels of normal composts such as Grade B.
- 30% *Gliricidia* amendment of normal compost at an application rate of 15 t/ha optimized vegetable production, particularly when used in nutrient-depleted sandy soils

REFERENCES

1. Singh, A. (1974) Use of organic material and green manures as fertilizers in developing countries. *Soil Bulletin* 27.
2. Wilson, G.F., Kang, B.T., and Mulongoy, K. (1986) Alley crop-ping trees as source of green manure and mulch in the tropics.*Biol. Agric. Hortic.* **3**, 251–267.
3. Agboola, A.A., Wilson, G.F., Getahum, A., and Yamoah, C.F. (1981) *Gliricidia sepium*: a shrub legume with a future for con- tinuous arable cropping in the humid tropics. Paper presented at the Workshop on Agroforestry in the African Humid Tropics, University of Ibadan, Ibadan, Nigeria, April 27–May 1.
4. Yamoah, C.F., Agboola, A.A., and Mulongoy, K. (1986) Decom- position and weed control by pruning of selected alley cropping shrubs. *Agrofor. Syst.* **4(3)**, 239–246.
5. Handayanto, E., Giller, K.E., and Cadish, G. (1997) Regulating N release from legume tree prunings by mixing residues of dif- ferent quality. *Soil Biol. Biochem.* **29(9–10)**, 1417–1426.
6. Agboola, A.A. and Obigbesan, C.O. (1974) Response of maize, rice and cassava to fertilizers in western Nigeria. First National Seminar on Planning and Fertilizer Use Development in Nigeria (Mines).
7. Egawa, T. (1975) Utilization of organic materials as fertilizers in
8. Japan. In Organic Materials as Fertilizers. FAO Soils Bulletin
9. 27. Food and Agriculture Organization, Rome. pp. 253–270.
10. Cooke, C.W. (1982) *Fertilizing for Maximum Yield.* Granada, London.
11. Peverly, J.H. and Gates. P.B. (1994) Utilization of municipal solid waste and sludge composts in crop production systems in sew- age sludge: land utilization and the environment. Proceedings of the Conference of the

American Society of Agronomy, Madison, WI. August 11–12, 1993.

12. Duncan, A. (1975) Organic materials as fertilizers. In Economic aspects of the use of organic materials as fertilizers. FAO/SIDA Expert Consultation, Rome, December 2–6, 1974. In Organic Materials as Fertilizers. FAO Soils Bulletin 27. Food and Agri- culture Organization, Rome.
13. Araji, A.A. and Stodick, L.D. (1990) Economic potential of feed-lot waste utilization in agricultural production. *Biol. Waste.* **32(2)**, 111–124.
14. Bouyoucus, C.H. (1951) A recalibration of the hydrometer method for making mechanical analysis of soils. *Agron. J.* **43**, 434–438.
15. Peech, M. (1965) Hydrogen-ion activity. In *Methods of Soil Analy- sis. Part 2. Chemical and Microbiological Properties*. Black, C.A, Ed. Agronomy Monograph No. 9. American Society of Agronomy, Madison, WI. pp. 914–926.
16. Black, C.A., Ed. (1965) Methods of Soil Analysis. Part 2. Chem*ical and Microbiological Properties.* Agronomy Monograph No
17. Bray, R.H. and Kurtz, O. (1945) Determination of total organic and available forms of phosphorus. *Soil Sci.* **59**, 45–49.
18. Tisdale, S. L., Nelson, W.L., and Beaton, J.D. (1985) *Soil Fertil- ity and Fertilizers.* 4th ed. Macmillan, New York.
19. Sillanpaa, M. (1982) Micronutrients and the Nutrient Status of Soils: A Global Study. FAO Soils Bulletin 48. Food and Agricul- ture Organization, Rome.
20. Aubert, H. and Pinta, M. (1977) *Trace Elements in Soils*. Elsevier, Amsterdam.
21. Wagner, S.F. (1992) *Introduction to Statistics*. HarperPerennial, New York, 369 p.
22. Norman, J.C. (1992) *Tropical Vegetable Crops*. Arthur H. Stockwell, Ilfracombe, U.K.
23. Mafongoya, P.L., Nair, P.K.R., and Dzowela, B.H. (1998) Min- eralization of nitrogen from decomposing leaves of multipurpose trees as affected by their chemical composition. *Biol. Fert. Soils* **27(2)**, 143–148.
24. Lionel, I., Wood, C.W., and Shannon, D.A. (2000) Decomposi- tion and Nitrogen release of prunings from hedgerow species assessed for alley cropping in Haiti. *Agron. J.* **92**, 501–511

Chapter 11

APPLICATION OF RURAL SLAUGHTERHOUSE WASTE AS AN ORGANIC FERTILIZER FOR POT CULTIVATION OF SOLANACEOUS VEGETABLES IN INDIA

Malancha Roy[1], Sukalpa Karmakar[1], Anupam Debsarcar[2], Pradip K Sen[3] and Joydeep Mukherjee[1]*

[1]School of Environmental Studies, Jadavpur University, Kolkata 700 032, India

INTRODUCTION

Globally, blood and rumen contents are the major abattoir wastes. In developing countries, a high proportion of the blood obtained from slaughtered animals ends as waste due to the lack of facilities for drying to produce blood meal. Industrial processing of blood by drum, flash, or spray drying produces good-quality blood meal but requires high capital outlay on heavy equipment as well as central slaughtering and collection of blood.

Rumen contents have about 85 g/kg of water content and constitute a disposal problem as their processing involves high investment and operating costs. Methods reported for processing rumen contents include field spreading or landfill, ensiling, drying using gas-fired rotary dryers, fluid-bed dryers, solar dryers, or by pressing to reduce the water content. These methods are not yet feasible in developing countries where slaughtering is mainly done in small and scattered units, which makes collection of large quantities of blood and rumen contents difficult. In addition, electricity and water supply are either lacking or irregular in most developing countries.

Therefore, the evolution of acceptable processing technologies for abattoir wastes is important (Makinde and Sonaiya 2010).

In a low-cost recycling method for abattoir wastes, a blend of bovine blood and rumen digesta, bovineblood- rumen-digesta-mixture (BBRDM), has been utilize as a replacement to full-fat soybean meal in broiler chickens' starter and finisher diets (Odunsi et al. 2004).

BBRDM was used as a replacement for groundnut cake and fishmeal in the diets of layer chickens (Odunsi 2003). In the present study, the application of BBRDM as a fertilizer and soil conditioner is being attempted for the first time. Generally, solanaceous vegetables require a large quantity of major nutrients like nitrogen (N), phosphorus (P), and potassium (K) for better growth and fruit yield. It is impractical to apply expensive fertilizer inputs for crops of marginal returns, and the rising cost of inorganic fertilizers have made them out of reach of small farmers in India. The application of

BBRDM to solanaceous vegetables could be an attractive proposal. This study was conducted in the perspective of Magrahat village, South 24 Parganas district of

West Bengal state, India, where the existing practice of random application of local slaughterhouse wastes as fertilizer was endeavored to be scientifically investigated.

The plants cultivated in this study were tomato, brinjal, and chili. Methods

Characterization of blood and rumen digesta Bovine blood and rumen digesta were collected from freshly annihilated animals. The following parameters in waste bovine blood were measured: pH by electrometric method, total solids (TS) by gravimetric method (103°C to 105°C), total suspended solids (TSS) by gravimetric method (103°C to 105°C), soluble 5-day biochemical oxygen demand (BOD5) at 20°C, chemical oxygen demand (COD) by open reflux method, total phosphorus (TP) by vanadomolybdophosphoric acid colorimetric method, total Kjeldahl nitrogen (TKN) by macro-Kjeldahl method, oil and grease by partition gravimetric method, and potassium (K) by flame photometric method. The characterization of rumen digesta was also done by measuring pH, TS, TSS, oil and grease, soluble BOD5, COD, TKN, TP, and K. Samples were obtained on five separate days and determinations performed in triplicate sets.

Preparation of BBRDM

Fresh blood of cattle was collected in containers immediately following the annihilation of animals. Bovine blood and rumen digesta were weighed in ratios of 1:1, 2:1, and 3:1 in three containers. Containers were placed on a coal-fired earthen stove and boiled for 90 min. The mixture was constantly stirred until the content was substantially free of water. The mass was sun dried for 3 days to obtain BBRDM. The final product was a coarse granular powder and could be easily spread.

N, P, and K analysis of BBRDM

The N, P, and K contents of BBRDM (ratios of 1:1, 2:1, and 3:1) were determined as described in the subsection 'Characterization of blood and rumen digesta.' All experiments were performed thrice in duplicate sets.

N and P mineralization during the storage of BBRDM

Immediately after the preparation of BBRDM (ratios of 1:1, 2:1, and 3:1), available N and P were measured following Subbiah and Asija (1956) and Bray and Kurtz (1945), respectively. Further determinations were made at 5-dayintervals for 16 days during the drying of BBRDM. All experiments were performed thrice in duplicate sets.

Cultivation of plants and determination of the dose and application frequency of BBRDM

The plants cultivated in this study were tomato (Solanum lycopersicum var. Patharkuchi), brinjal (Solanum melongena var. Jhuribegun), and chili (Capsicum annuum var. Sonamukhi). Seedlings of approximately the same height, purchased from a local nursery, were planted, one seedling in one pot. Different tubs, (1)filled with soil (of village Magrahat), (2) soil + diammonium phosphate (DAP) (NH4)2HPO4 (N/P/ K = 18:46:0), (3) soil + BBRDM (composed of one part of blood and one part of rumen digesta, 1:1), (4) soil + BBRDM (composed of two parts of blood and one part of rumen digesta, 2:1),(5) soil + BBRDM (composed of three parts of blood and one part of rumen digesta, 3:1), were used for the plantation of tomato, chili, and brinjal, a total of eight tubs for each of the five treatments following a completely randomized design. In the first series of experiments, BBRDM was applied at different doses (2.5, 5.0, and 10.0 g/kg of soil) at the time of planting. In the second series of experiments, the doses were applied to the plants on second (after an initial 15 days of plantation) and sixth weeks during a 20-week period of cultivation.

All experiments were performed thrice in duplicate sets. Plant growth parameters as mentioned in the 'Results' section (Tables 1, 2, and 3) were measured or counted every week to compare plant growth. Pot cultivation was done in the winter seasons (November to March) of 2009 to 2010, 2010 to 2011, and 2011 to 2012. The tubs were kept under a shed and watered every day. Pest and weed control was not necessary during the pot cultivation.

The mean maximum temperature was 27.4°C ± 2.2°C, and the mean minimum temperature was 16.4°C ±2.6°C. The fruits, as they matured, were picked toward the end of the cultivation.

Table 1 Comparative analysis of the growth parameters of tomato plants after 12 weeks

Plant growth parameters	**Soil treatments**				
	Soil	**Soil + DAP**	**Soil + BBRDM (1:1)**	**Soil + BBRDM (2:1)**	**Soil + BBRDM (3:1)**
Plant height (cm)	37.05 ± 0.5^{b}	40.5 ± 1^{ab}	42 ± 1.5^{ab}	42.5 ± 0.5^{a}	42 ± 1^{ab}
Number of leaves	10 ± 1^{b}	11 ± 1^{b}	18 ± 1^{a}	20 ± 2^{a}	18 ± 2^{a}
Leaf surface area (cm^2)	125.5 ± 1^{d}	139 ± 2.5^{c}	141 ± 1^{c}	154 ± 2^{a}	150 ± 2^{b}
Stem surface area (cm^2)	0.48 ± 0.05^{d}	0.66 ± 0.02^{c}	1.11 ± 0.04^{b}	1.43 ± 0.02^{a}	0.48 ± 0.02^{d}
Number of buds	22 ± 3^{d}	27 ± 3^{c}	34 ± 4^{b}	38 ± 2^{ab}	40 ± 2^{a}
Number of flowers	15 ± 1^{c}	18 ± 2^{c}	29 ± 1^{b}	28 ± 1^{b}	34 ± 2^{a}
Number of fruits	10 ± 1^{c}	10 ± 1^{c}	14 ± 1^{b}	19 ± 2^{a}	21 ± 3^{a}
Total fruit weight (g)	196.25 ± 3.46^{e}	337.64 ± 4.56^{d}	414.65 ± 3.06^{c}	719.77 ± 4.65^{b}	775.95 ± 4.58^{a}

Determination of N, P, and organic C content of soil

Available N (Subbiah and Asija 1956), available P (Bray and Kurtz 1945), and organic carbon (C) (Walkely and

Black 1934) were measured before and during cultivation in each tub every 2 weeks. All experiments were performed thrice in duplicate sets.

Microbiological analysis of soil

Soil samples of the 2nd, 8th, and 14th weeks (the fertilizers were applied in the second and sixth weeks) were considered for microbiological analysis. Rhizosphere soil was collected at a depth of 3 to 5 cm. Nutrient agar (NA) media were used for enumeration of the total bacterial population. Pikovskaya (PVK) medium was used for the enumeration of phosphate-solubilizing bacteria.

Ashbys mannitol (AM) agar medium was used for the selection of Azotobacter species (main nitrogen-fixing bacteria) that utilize mannitol and atmospheric nitrogen as sources of C and N, respectively. BG-11 broth was used for the growth of freshwater cyanobacteria. Cooke

Rose Bengal (CRB) agar base medium was used for the selective isolation of fungi. For compositions of NA, PVK, AM, BG-11, and CRB media, please see Additional file 1. Test and control soil samples (0.1 g) were taken and mixed with 10 ml of sterile water. Initial abundances of each type of microorganism in all the pots (at day 0) were recorded. Soil sample dilutions were vortexed for 5 min, and 100 µl of each dilution was plated on four types of media (NA, PVK, AM, and CRB) and inoculated in BG-11 liquid broth. NA and PVK plates were observed after incubation for 24 h at 37°C, AM plates after 48-h incubation at 37°C, and fungal plates after 72-h incubation at 25°C. Cyanobacteria were incubated under the light of 50-µmol photons m2/s intensity, 12 h a day for 4 weeks. Thereafter, the cyanobacterial mass was centrifuged and extracted with methanol, and the absorbance of chlorophyll a was measured at 663 nm (Pramanik et al. 2011). All experiments were done thrice in duplicate sets.

Appearance, taste, and food quality analysis

The mature fruits were visually observed and tasted by five individuals. The vegetables obtained from the plants were analyzed for total protein (Lowry's method), total carbohydrate (phenol sulphuric acid method), and fat (Soxhlet method), and the values were compared with vegetables grown with DAP. All experiments were done thrice in duplicate sets.

Table 2 Comparative analysis of the growth parameters of brinjal plants after 12 weeks

Plant growth parameters	Soil treatments				
	Soil	Soil + DAP	Soil + BBRDM (1:1)	Soil + BBRDM (2:1)	Soil + BBRDM (3:1)
Plant height (cm)	22.5 ± 0.5^{b}	36 ± 1^{a}	32.5 ± 1.5^{a}	32 ± 0.5^{a}	32.5 ± 0.5^{a}
Number of leaves	11 ± 1^{bc}	13 ± 2^{ab}	10 ± 1^{b}	15 ± 1^{a}	13 ± 1^{ac}
Leaf surface area (cm^{2})	136 ± 1.01^{c}	135 ± 1.06^{c}	163 ± 2.02^{a}	153 ± 2.25^{b}	164 ± 2^{a}
Stem surface area (cm^{2})	0.36 ± 0.02^{c}	0.35 ± 0.01^{c}	0.63 ± 0.01^{a}	0.53 ± 0.01^{b}	0.64 ± 0.02^{a}
Number of buds	6 ± 1.58^{c}	7 ± 1.56^{c}	8 ± 1.58^{bc}	10 ± 2.07^{ab}	13 ± 2.35^{a}
Number of flowers	3 ± 1.21^{c}	4 ± 1.2^{c}	6 ± 1.32^{bc}	8 ± 1.56^{ab}	11 ± 2.11^{a}
Number of fruits	1 ± 0.84^{d}	2 ± 1.67^{cd}	4 ± 1.19^{bc}	5 ± 1.23^{ab}	7 ± 1.58^{a}
Total fruit weight (g)	44.87 ± 2.99^{e}	149.3 ± 1.58^{d}	326.37 ± 5.6^{c}	420.61 ± 6.2^{b}	556.18 ± 7.89^{a}

Table 3 Comparative analysis of the growth parameters of chili plants after 12 weeks

Plant growth parameters	Soil treatments				
	Soil	Soil + DAP	Soil + BBRDM (1:1)	Soil + BBRDM (2:1)	Soil + BBRDM (3:1)
Plant height (cm)	32 ± 2^{b}	29 ± 2^{c}	42 ± 2^{a}	41.5 ± 1^{a}	42.5 ± 1^{a}
Number of leaves	108 ± 4^{e}	129 ± 3^{d}	150 ± 6^{c}	155 ± 3^{b}	180 ± 4^{a}
Leaf surface area (cm^{2})	12.7 ± 0.7^{b}	11.4 ± 1.2^{d}	11.9 ± 1.5^{c}	13.8 ± 1.3^{a}	14 ± 1.5^{a}
Stem surface area (cm^{2})	0.41 ± 0.02^{d}	0.43 ± 0.02^{d}	1.74 ± 0.02^{a}	1.15 ± 0.01^{b}	1.09 ± 0.01^{c}
Number of buds	12 ± 1.68^{e}	28 ± 2.03^{c}	25 ± 1.99^{d}	32 ± 2.5^{b}	46 ± 3.78^{a}
Number of flowers	10 ± 2.07^{e}	22 ± 1.98^{c}	17 ± 1.63^{d}	28 ± 2.02^{b}	39 ± 2.95^{a}
Number of fruits	7 ± 1.46^{e}	18 ± 1.56^{c}	10 ± 1.45^{d}	25 ± 2.14^{b}	36 ± 2.88^{a}
Total fruit weight (g)	6.79 ± 1.39^{e}	28.11 ± 2.03^{c}	13.39 ± 2.15^{d}	64.39 ± 3.78^{b}	100.81 ± 4.62^{a}

Statistical methods

Analysis of variance (ANOVA) for one-way classified data on plant growth parameters was performed using SPSS for Windows version 11.5 (SPSS Inc., Chicago, IL, USA).

Differences were significant at 0.05 level. Tukey's test was performed as post hoc analysis of one-way ANOVA for pairwise comparison. Arithmetic mean of the microbial numbers and mass (in the case of cyanobacteria) was determined considering the values at the 2nd, 8th, and 14th weeks. Standard deviations of the means were calculated. Tukey's test was done as described before.

Results and discussion

Characterization of waste blood and rumen digesta The average pH of blood was 8.0 ± 0.1, the BOD5 was 69,558 ± 880 mg/l, the COD was 268,571 ± 3194.5 mg/l, and TKN was 12,414 ± 721.3 mg/l. The concentrations of oil and grease, TSS, TS, TP, and K were 27.3 ± 1.5, 433,192 ± 2,476, 824,762 ± 4,752, 236 ± 17, and 3,933 ± 21 mg/l, respectively. The N, P, and K content of rumen digesta was determined as N/P/K = 945.0 ± 30:1,817.33 ± 11.24:400 ± 14 mg/kg. The average pH of rumen digesta was 8.0 ± 0.1, soluble BOD5 was 120 ± 70 mg/l, the COD was 3,440 ± 27 mg/l, TS was 57,400 ± 437 mg/l, TSS was 42,000 ± 527 mg/l, while oil and grease was 77,100 ± 657 mg/l. Error ranges indicate one standard deviation from the mean (n = 15). These parameters characterized the slaughterhouse refuse as highly polluting wastes.

Characterization of BBRDM

The N, P, and K, content of BBRDM was 49,440.64 ± 236.25:1,628.75 ± 322.25:9,367.94±1,132.06 mg/kg. The N/P/K ratio was approximately 30.36:1:5.75. Error ranges indicate one standard deviation from the mean (n = 6).

Plant growth and yield of vegetables

Results of the experiments done as described in the subsection 'Cultivation of plants and determination of the dose and application frequency of BBRDM' in the 'Methods section showed that all plants treated with BBRDM (3:1) died when 2.5 g BBRDM/kg of soil was added at the time of planting. When fertilized with 5.0 g BBRDM/ kg of soil at the time of planting, all plants treated with BBRDM

(3:1) and half of the plants treated with BBRDM (2:1) died. When fertilized with 10 g BBRDM/kg of soil at the time of planting, all plants treated with the three preparations of BBRDM died. On the other hand, none of the plants died when BBRDM was applied at the second and sixth weeks, and 5.0 g of BBRDM/kg soil produced the best results compared to 2.5 g as well as 10.0 g BBRDM/kg of soil in terms of plant growth parameters and yields of fruits.

Results are shown in Tables 1, 2, and 3. Error ranges in the tables indicate one standard deviation from the mean (n = 24). Table 1 shows the results of the effect of BBRDM on the growth of tomato plants.

Tukey's test shows that the highest yield in terms of total fruit weight was obtained when BBRDM (3:1) was applied. In terms of the number of fruits, BBRDM (3:1) and BBRDM (2:1) produced more yields than the other treatments, though the difference between BBRDM (3:1) and BBRDM (2:1) is not statistically significant. The onset of flowering and fruiting was observed to be 2 weeks earlier than DAP. Chalkos et al. (2010) observed in pot studies that Mentha spicata compost added at 4% to 8% (w/w) proved to be a very promising soil amendment.

Results obtained in pot studies showed that anaerobically fermented pig slurry could be a suitable alternative to the use of mineral fertilizer for tomato cultivation (Kouřimská et al. 2009). Experiments carried out in pots revealed that the yield of tomato and critical nutritional parameters showed significant increase upon the application of biowaste from the tobacco industry (Chaturvedi et al. 2008).

Table 2 shows the results of the effect of different soil conditioners on the growth of brinjal plants. The highest yield, as recorded by total fruit weight, was attained in soils treated with BBRDM (3:1). In terms of the number of fruits, BBRDM (3:1) and BBRDM (2:1) produced higher yields than the other applications, though the difference between BBRDM (3:1) and BBRDM (2:1) is not statistically significant nor is the difference between

BBRDM (2:1) and BBRDM (1:1). The onset of flowering and fruiting was observed to be 2 weeks earlier than in plants treated with inorganic fertilizer. Górecki and

Górecki (2010) reported that sheep wool could serve as a valuable and environmentally friendly fertilizer for brinjal cultivation in pot studies.

Table 3 shows the results of the effect of BBRDM on the growth of chili plants. Tukey's test shows that the highest yields in terms of the number of fruits and total fruit weight were obtained when BBRDM (3:1) was applied.

The onset of flowering and fruiting was observed to be 2 weeks earlier than in plants treated with DAP. Damke et al. (1988) and Surlekov and Rankov (1989) noted the enhancement of height and number of branches and leaves of chili plants upon the application of farmyard manure with inorganic supplements.

The induction of early flowering is probably due to better nutritional status of the plants. Increased production of leaves might have helped to elaborate more photosynthates and induce flowering, thus effecting to early initiation of the flower bud. The results are in conformity with the report for tomato (Sharma and Mahendra 1963) as well as for chili (Revanappa et al. 1998; Patil and Biradar 2001). The increase in leaf area may be attributed to more number of leaves per plant. Increased plant height may be due to the increased uptake of primary nutrients, which might have enhanced cell division and cell elongation. Yield is the manifestation of morphological, physiological, biochemical, and growth parameters and is considered to be the result of efficient trapping and conversion of solar energy. This might have helped in producing a higher amount of carbohydrates which might have translocated from the source (leaf) to the reproductive parts (sink), resulting in more number of fruits and fruit weight. Yield attributes are closely associated with growth components like plant height, number of leaves, and leaf area per plant (Jagadeesha 2008). Higher number of fruits and fruit weight obtained in plants grown with BBRDM may be due to the increased growth components of the three crops. Similar findings were also reported in brinjal by Amburani and Manivannan (2002).

P was lower than that of N. Vegetable crops require an adequate and continuous supply of N for proper growth and productivity. The presence of N in plant tissue is known to increase its vegetative growth and chlorophyll formation.

Nitrogen is the chief constituent of protein that leads to cell enlargement, cell division, and ultimately increase plant growth.

Phosphorus is known to increase root growth, seed formation, and flower development.

In our experiments, N and P are obtained from BBRDM. Even if a proportion of soil nitrogen is fixed from the air, this is not sufficient to maintain agricultural productivity and requires external addition of fertilizers. In our experiments, the yields obtained from plants without added fertilizers were significantly lower than the yields obtained upon the addition of DAP or BBRDM. Approximately two and four times more yields of tomato were attained upon the addition of DAP and BBRDM (3:1), respectively. The addition of DAP and BBRDM (3:1) improved the yields of brinjal by about 4 and 15 times, respectively. Similarly, the application of DAP and BBRDM (3:1) improved the yields of chili by 3 and 12 times, respectively. BBRDM promoted the growth of nitrogen-fixing bacteria that may have contributed to the soil N content. However, the distinction between direct N derived from BBRDM and that gained from nitrogen-fixing bacteria was not possible in this study and can be objectives of future fundamental studies on soil chemistry.

In general, solanaceous vegetables require more amounts of nitrogen compared to phosphorus. For tomato, to produce 1 metric ton of fresh fruit, plants need to absorb, on average, 2.5 to 3 kg N and 0.2 to 0.3 kg P. Brinjals need 3 to 3.5 kg N and 0.2 to 0.3 kg P, while chilies need 3 to 3.5 kg N and 0.8 to 1 kg P (Hegde 1997). From Figure 1a,b,c, it is observed that the phosphorus content of soils treated with BBRDM progressively increases, indicating accumulation.

In our experiments, phosphorus mineralization using soil microbes continued following the application, and due to the low uptake, accretion occurred. Mohammadi et al. (2009) observed that the mineralization of organic P may release P into the soil solution, contributing to the observed high water-soluble P content with a high-P-manured soil. Gaskell et al. (2006) also noted that soil phosphorus can rapidly build up to high levels when composts and other organic amendments are used.

One important characteristic of an agricultural waste product to be used as an organic fertilizer is the potentially available N (Cavaleri et al. 2004). Nitrogen content in BBRDM (5.5%, w/w) was similar to cotton meal (8.21%) and castor oil (7.54%) and higher than traditional waste products used as fertilizers like sugarcane bagasse (0.24%), wood ash (0.51%), and bovine manure (0.77%) (Lima et al. 2011). The majority of the N present in the soil is found in the organic form, which is not available for plant absorption. To be taken up by theplants, N has to be in the inorganic form, like nitrate or ammonia (Savvas et al. 2010). Some organic residues are slowly mineralized due to low N content. When they are added to the soil, N and P contents can be temporarily reduced due to microorganism immobilization (López-Piñeiro et al. 2008). In those materials, N is slowly released, and the residual effect is expected to last longer (Cabrera et al. 2005). BBRDM, on the contrary, was mineralized extremely fast, as seen in Figure 2.

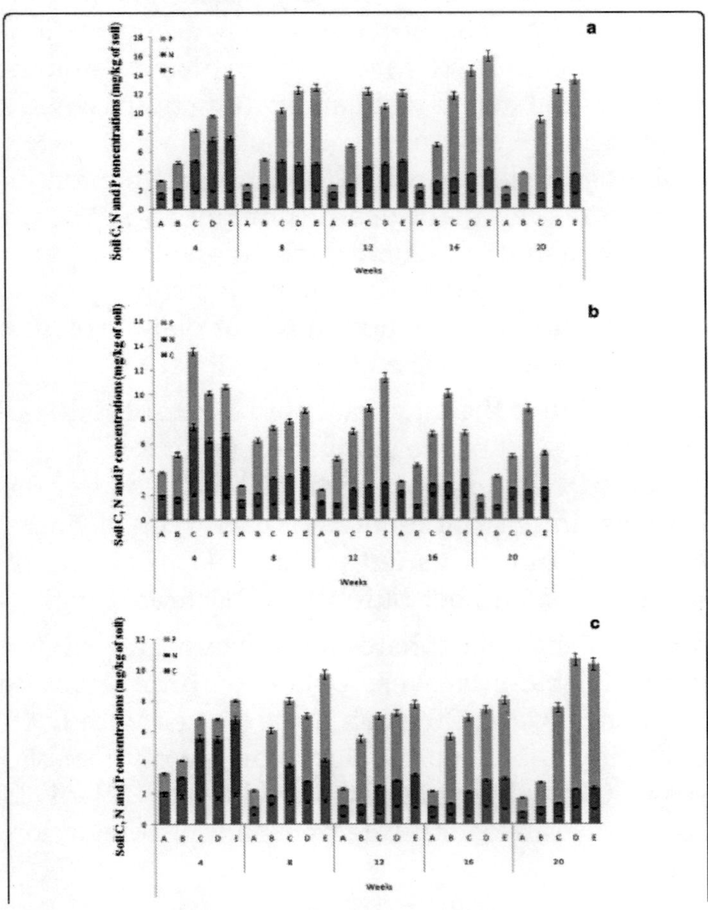

Figure 1 Soil C, N, and P concentrations during cultivation of (a) tomato, (b) brinjal, and (c) chili.

The availability of N was considerably higher in BBRDM (3:1) compared with the other two ratios. No significant differences in available P were observed among the three preparations of BBRDM (Figure 2). Available N and available P increased significantly on the sixth day of drying of

BBRDM and subsequently leveled off. The high standard deviations may arise from variations during weighing as BBRDM is a coarse granular powder. A large quantity of

N was released in a short period of time; high concentrations of mineral N (nitrate and ammonia) caused plant toxicity when applied at the time of planting to very young plants. Phytotoxicity caused by excess of N usually occurs due to accumulation of NH4 + (Savvas et al. 2010). As BBRDM was rich in N and P and mineralized fast, this material enhanced yield and productivity when applied to mature plants after 15 days of plantation. Blends of castor meal and husks containing higher than 4.5% N caused reduction in castor oil plant growth and even plant death (Lima et al. 2011). Castor meal was used as an organic fertilizer for wheat plants (Gupta et al. 2004). When the meal was added to the soil on the sowing day, increased mineral-N doses caused reduction in wheat plant growth because N content in the soil was increased to a toxic level.

Characterization of the microbial population of soil The fertilizers were added at the start of the cultivation (after 2 weeks) and after 6 weeks. Therefore, microbial counts were obtained after the 2nd, 8th, and 14th weeks. Initial abundances of each type of microorganism in all the pots (at day 0) varied within 5%. In all the three crops cultivated, in general, higher numbers of bacteria, fungi,

Azotobacter, and phosphate solubilizers and higher amount of cyanobacterial biomass were obtained from soils treated with BBRDM (3:1) or BBRDM (2:1) than those treated with DAP with one exception. During tomato cultivation, a higher number of phosphate solubilizers were obtained in soils treated with BBRDM (1:1) (Tables 4, 5, and 6). Error ranges indicate one standard deviation from the mean (n = 6).

Phosphorus is an essential component of the energy compounds (ATP and ADP) and phosphoproteins. Phosphate solubilizers liberate P in soil and make it available for plants. Phosphate-solubilizing bacteria release organic and inorganic acids which reduce soil pH leading to change of P and other nutrients to available forms ready for uptake

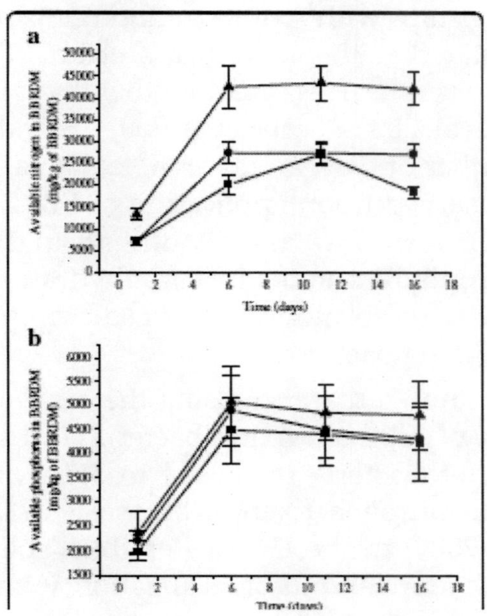

Figure 2 Available (a) nitrogen and (b) phosphorus during sun drying of BBRDM after preparation

Table 4 Soil microbial abundance during tomato cultivation

Number of microorganisms	Soil treatments				
	Soil	Soil + DAP	Soil + BBRDM (1:1)	Soil + BBRDM (2:1)	Soil + BBRDM (3:1)
Fungi ($\times 10^3$/ml) in CRB	6.44 ± 1.13[b]	7.83 ± 1.5[b]	9.39 ± 1.6[b]	14.33 ± 2.64[a]	14.44 ± 2.5[a]
Bacteria ($\times 10^7$/ml) in NA	23.99 ± 1.69[d]	29.33 ± 1.56[c]	29.67 ± 1.62[c]	35.61 ± 2.11[b]	45.67 ± 2.93[a]
Azotobacter ($\times 10^7$/ml) in AM	4.11 ± 0.66[c]	5 ± 1.19[c]	19.61 ± 0.71[b]	26.5 ± 1.13[a]	25.78 ± 2.28[a]
Phosphate solubilizers ($\times 10^7$/ml) in PVK	5.83 ± 0.75[e]	11.5 ± 0.91[d]	19.17 ± 0.62[a]	13.5 ± 0.62[c]	17.5 ± 0.62[b]
Chlorophyll *a* content (µg/g of soil)	0.49 ± 0.04[c]	1.68 ± 0.07[b]	2.13 ± 0.36[ab]	2.2 ± 0.47[a]	2.49 ± 0.14[a]

by plants. Furthermore, phosphobacteria augment plant growth through the biosynthesis of growth-promoting substances like vitamin B2 and auxins. The results of the present investigation are in confirmation with the findings in brinjal (Nanthakumar and Veeraraghavathatham 2000), in chili (Govindarajan and Thangaraju 2001), and in tomato (Kumaran et al. 1998; Reddy 1999). Nitrogen-fixing microbes are known to improve the growth and productivity of crops. Azotobacter species are free-living nitrogenfixing bacteria and normally fix molecular nitrogen from the atmosphere without symbiotic relationship with plants, although some Azotobacter species are associated with plants (Kass et al. 1971). In addition, Azotobacter also synthesizes auxins, thereby stimulating plant growth (Ahmad et al. 2005; Rajaee et al. 2007). Significant increase in plant height, number of leaves, leaf area, leaf area index, and early flowering was observed upon application of BBRDM. This may be attributed to the increased uptake of nutrients in the plants leading to enhanced chlorophyll content and carbohydrate synthesis as well as increased activity of hormones produced by nitrogen-fixing and phosphate-solubilizing bacteria.

It is generally known that increasing the concentration of soil N inhibits the effect of nitrogen-fixing bacteria. In the present study, the number of azotobacteria increased in spite of increasing soil N content. Would nitrogen fixation be affected? In the study by Mikanová et al. (2009) in the Czech Republic, soils were sampled from the plots with four variants of fertilization: without fertilization, mineral fertilization, farmyard manure, and farmyard manure with mineral fertilization. Counts of Azotobacter spp. and potential nitrogenase activity were very closely related to the nitrogen content in soil. According to the authors, nitrogen fertilization in organic form (farmyard manure) increased the counts of Azotobacter spp. and subsequently also the potential nitrogenase activity. Inorganic (mineral) fertilization had no effect on the measured characteristics.

Their results were in accordance with those published by Czakó et al. (2007), where the authors reported significantly increased counts of Azotobacter spp. in dump substrate upon the addition of 400 metric tons of compost per hectare during a recultivation study conducted in the

North Bohemian coal basin, Czech Republic. It was concluded that soil conditions, especially a sufficient supply of organic matter that increases the supply of nitrogen to soil, are suitable for the development of Azotobacter spp. (Mikanová et al. 2009). Significant correlation between the number of Azotobacter spp. and the contents of organic C and total N in Polish soils indicated that soil fertility was an important factor influencing colonization of soils by Azotobacter

(Martyniuk and Martyniuk 2003). There was a significant correlation between soil pH, total nitrogen, and organic carbon with Azotobacter population as reported by Mazinani et al. (2012) in their study on Iranian soils. The number of bacteria in unit gram of soil increased with increasing concentrations of C and N.

Number of microorganisms	Soil treatments				
	Soil	Soil + DAP	Soil + BBRDM (1:1)	Soil + BBRDM (2:1)	Soil + BBRDM (3:1)
Fungi ($\times 10^3$/ml) in CRB	9.28 ± 1.1^c	12.17 ± 0.81^b	17.56 ± 1.89^a	19 ± 1.96^a	17.56 ± 1.96^a
Bacteria ($\times 10^7$/ml) in NA	20.22 ± 1.69^c	59.39 ± 1.83^b	60.45 ± 2.4^b	85 ± 2.9^a	86.05 ± 2.23^a
Azotobacter ($\times 10^7$/ml) in AM	2.5 ± 0.62^e	6.11 ± 0.72^d	13.11 ± 0.72^c	17.72 ± 1.42^b	20.39 ± 1.36^a
Phosphate solubilizers ($\times 10^7$/ml) in PVK	7.06 ± 0.88^e	15.22 ± 1.19^d	21.33 ± 1.3^c	34.67 ± 1.82^b	44.45 ± 1.86^a
Chlorophyll *a* content (μg/g of soil)	0.39 ± 0.02^e	1.06 ± 0.02^d	1.76 ± 0.03^b	1.56 ± 0.04^c	3.14 ± 0.03^a

Although Azotobacter, in general, is a nitrogen fixer, the addition of nitrogen (as NO3 – as well as NH4 +) in the medium decreased the lag phase and generation time and thus increased the number of the bacteria (Vermani et al. 1997). Results of another investigation showed that the addition of NH4Cl and NaNO3 into the medium increased the growth of 1 wild and 13 mutant types of Azotobacter vinelandii (Iwahashi and Someyo 1992). Laane et al. (1980) presented evidence on the direct depressing effect of ammonium chloride on nitrogen fixation by A. vinelandii and attributed it to the inhibition of the electron transport system to nitrogenase. Abd-el-Malek et al. (1975), on the contrary, observed no relationship between NH4 +-N concentration and densities of Azotobacter in the Nile valley soils of Egypt, indicating that the presence of NH4 +-N had no effect on the multiplication of the organism. The absence or presence of NH4 +-N did not affect the rate of sugar consumption by Azotobacter and corroborated the observation that NH4 +-N did not affect their multiplication rate. The authors further showed that the addition of composted organic matter increased the efficiency of N2fixation and prevented the competitive inhibition of ammonium salts.

This could be attributed to the following factors: (a) soil organic matter possesses the property of cation exchange and its presence in a medium results in it adsorbing the

NH4 +-N, rendering them less available to Azotobacter than N2, and (b) the presence of certain growth factors or trace elements which may increase the efficiency of fixation in addition to enabling the organism to overcome the unfavorable effect of ammonium salts on fixation.

In the light of these literature reports, the question of finding increased azotobacteria with increased N concentration as well as the putative relationship between soil nitrogen concentration and nitrogen fixation as apparent through the present study should be addressed as a separate detailed investigation. Further experiments on the determination of non-symbiotic nitrogen fixation by measurement of ethylene gas and nitrogenase activity are necessary to arrive at a definite conclusion.

Soil fungi perform vital functions within the soil in relation to nutrient cycling. It is now widely recognized that decomposition, mineralization, and plant-fungal mutualisms regulate the movement of soil-borne elements into the plant roots. Soil fungi are also responsible for disease containment and water dynamics; all of which help plants become healthy and robust (Christensen 1989). Soil cyanobacteria release secondary metabolites, which can be mineralized using the microflora and are thus beneficial to agricultural crops. These substances may be growth promoters and/or inhibitors for soil microflora. Cyanobacteria play an important role in soil chemical transformations and thus influence the bioavailability of major nutrients like P (Singh et al. 2011).

Organic fertilizers increase the numbers of beneficial microbes in the soil that contribute to soil fertility. For tomato cultivation, Tu et al. (2006) showed that microbial biomass and microbial activity were generally higher in organically managed soils, cotton gin trash being most effective, than conventional soil treatment with synthetic fertilizers. Similarly, cotton gin trash enhanced the populations of beneficial soil microbes during tomato cultivation as reported by Bulluck and Ristaino (2002). Microbial populations found at the rhizosphere, rhizoplane, and bulk soil due to the application of compost-based fertilizer, singly or in combination with urea, increased two- to fivefold relative to treatment with synthetic NPK fertilizer during cultivation of tomatoes (Taiwo et al. 2007).

Appearance, taste, and food quality

Normal color of the fruits was attained, as may be seen in Additional file 2: Figure S1a,b,c. The taste of the vegetables in terms of sourness, sweetness, and hotness as reported by five individuals was identical to normal tomato, brinjal, and chili. The composition of vegetables obtained from cultivation with BBRDM (3:1) (per 100 g): tomato-carbohydrate 3.89 ± 0.45 g, fat 0.19 ± 0.04, protein 1 ± 0.2 g per; brinjal-carbohydrate 5.5 ± 0.4 g, fat 0.18 ± 0.03 g, protein 0.82 ± 0.02 g; chili-carbohydrate 8 ± 0.3 g, fat 0.39 ± 0.02 g, protein 2.1 ± 0.1 g. The composition of vegetables obtained from cultivation with DAP (per 100 g): tomato-carbohydrate 3.9 ± 0.4 g, fat 0.2 ± 0.01 g, protein 0.9 ± 0.05 g; brinjal-carbohydrate 5.7 ± 0.6 g, fat 0.19 ± 0.01 g, protein 1.01 ± 0.02 g, and chili-carbohydrate 8.8 ± 0.4 g, fat 0.4 ± 0.01 g protein 1.9 ± 0.02 g. Thus, the vegetables grown with BBRDM and DAP were of comparable quality.

CONCLUSIONS

Highly polluting slaughterhouse could be gainfully utilized, which would promote the preservation of a healthy environment around rural slaughterhouses. Field investigations are underway, and preliminary results have shown similar trends as observed in this pot study.

REFERENCES

1. Abd-el-Malek Y, Hosny I, Shawky BT (1975) Availability of soil ammoniacal nitrogen to Azotobacter and its effect on nitrogen fixation. Zbl Bakt Abt II Bd130:584–597
2. Ahmad F, Ahmad I, Khan MS (2005) Indole acetic acid production by the indigenous isolates of Azotobacter and fluorescent Pseudomonas in the presence and absence of tryptophan. Turk J Biol 29:29–34
3. Amburani A, Manivannan K (2002) Effect of integrated nutrient management on growth in brinjal (Solanum melongena L.) cv. Annamalai. South Indian Hort 50:377–386
4. Bray RH, Kurtz LT (1945) Determination of total, organic, and available forms of phosphorus in soils. Soil Sci 59:39–45
5. Bulluck LR III, Ristaino JB (2002) Effect of synthetic and organic soil fertility amendments on southern blight, soil microbial communities, and yield of processing tomatoes. Phytopathology 92:181–189
6. Cabrera ML, Kissel DE, Vigil MF (2005) Nitrogen mineralization from organic residues: research opportunities. J Environ Qual 34:75–79
7. Cavaleri MA, Gilmore DW, Mozaffari M, Rosen CJ, Hallbach TR (2004) Hybrid poplar and forest soil response to municipal and industrial by-products: a greenhouse study. J Environ Qual 33:1055–1061

8. Chalkos D, Kadoglidou K, Karamanoli K, Fotiou C, Pavlatou-Ve AS, Eleftherohorinos IG, Constantinidou H-IA, Vokou D (2010) Mentha spicata and Salvia fruticosa composts as soil amendments in tomato cultivation. Plant Soil 332:495–509

9. Chaturvedi S, Upreti DK, Tandon DK, Sharma A, Dixit A (2008) Bio-waste from tobacco industry as tailored organic fertilizer for improving yields and nutritional values of tomato crop. J Environ Biol 29:759–763

10. Christensen M (1989) A view of fungal ecology. Mycologia 81:1–19

11. Czakó A, Mikanová O, Ustak S (2007) The effect of inoculation on reclaimed soils. Počvoved Agrochim 1:232–237

12. Damke MM, Kawarkhe VJ, Patil CO (1988) Effect of phosphorous and potassium on growth and yield of chilli. Punjabrao Krishi Vidhyapeeth Res J 12:110–114

13. Gaskell M, Smith R, Mitchell J, Koike ST, Fouche C, Hartz T, Horwath W, Jackson L (2006) Soil fertility management for organic crops. University of California,

14. Division of Agriculture and Natural Resources Publication, CA, p 7249 Górecki RS, Górecki MT (2010) Utilization of waste wool as substrate amendment in pot cultivation of tomato, sweet pepper, and eggplant. Pol J Environ Stud 19:1083–1087

15. Govindarajan K, Thangaraju M (2001) Azospirillum - a potential inoculant for horticulture crops. South Indian Hort 49:223–235

16. Gupta AP, Antil RS, Narwal RP (2004) Utilization of deoiled castor cake for crop production. Arch Agron Soil Sci 50:389–395

17. Hegde DM (1997) Nutrient requirements of solanaceous vegetable crops. Food and Fertilizer Technology Center Publication Database., http://www. agnet. org/library.php?func=view&id=20110801133428 (last accessed on 05/02/2013

18. Iwahashi H, Someyo J (1992) Oxygen sensitivity of nitrogenase is not always a limiting factor of growth under nitrogen-fixing conditions in Azotobacter vinelandii. Biotechnol Lett 4:227–232

19. Jagadeesha V (2008) Effect of organic manures and biofertilizers on growth, seed yield and quality in tomato (Lycopersicon esculentum Mill.) cv. Megha. Master of Science (Agriculture) thesis. University of Agricultural Sciences, Dharwad Kass DL, Drosdoff M, Alexander M (1971)

20. Nitrogen fixation by Azotobacter paspali in association with bahiagrass (Paspalum notatum). Soil Sci Soc Am J 35:286–289

Chapter 12

VERMICOMPOSTING: RECYCLING WASTES INTO VALUABLE ORGANIC FERTILIZER

Nagavallemma KP, Wani SP, Stephane Lacroix, Padmaja VV, Vineela C, Babu Rao M and Sahrawat KL

2004. Vermicomposting: Recycling wastes into valuable organic fertilizer. Global Theme on Agrecosystems Report no. 8. Patancheru 502 324, Andhra Pradesh, India: International Crops Research Institute for the Semi-Arid Tropics. 20 pp.

BACKGROUND

Environmental degradation is a major threat confronting the world, and the rampant use of chemical fertilizers contributes largely to the deterioration of the environment through depletion of fossil fuels, generation of carbon dioxide (CO_2) and contamination of water resources. It leads to loss of soil fertility due to imbalanced use of fertilizers that has adversely impacted agricultural productivity and causes soil degradation. Now there is a growing realization that the adoption of ecological and sustainable farming practices can only reverse the declining trend in the global productivity and environment protection (Aveyard 1988, Wani and Lee 1992, Wani et al. 1995).

On one hand tropical soils are deficient in all necessary plant nutrients and on the other hand large quantities of such nutrients contained in domestic wastes and agricultural byproducts are wasted. It is estimated that in cities and rural areas of India nearly 700 million t organic waste is generated annually which is either burned or land filled (Bhiday 1994). Such large quantities of organic wastes generated also pose a problem for safe disposal. Most of these organic residues are burned currently or used as land fillings. In nature's laboratory there are a number of organisms (micro and macro) that have the ability to convert organic waste into valuable resources containing plant nutrients and organic matter, which are critical for maintaining soil productivity. Microorganisms and earthworms are important biological organisms helping nature to maintain nutrient flows from one system to another and also minimize environmental degradation. The earthworm population is about 8–10 times higher in uncultivated area. This clearly indicates that earthworm population decreases with soil degradation and thus can be used as a sensitive indicator of soil degradation. In this report a simple biotechnological process, which could provide a 'win-win' solution to tackle the problem of safe disposal of waste as well as the most needed plant nutrients for sustainable productivity is described (Wani 2002).

What is Vermicomposting?

Vermicomposting is a simple biotechnological process of composting, in which certain species of earthworms are used to enhance the process of waste conversion and produce a better end product. Vermicomposting differs from composting in several ways (Gandhi et al. 1997). It is a mesophilic process, utilizing microorganisms and earthworms that are active at 10–32°C (not ambient temperature but temperature within the pile of moist organic material). The process is faster than composting; because the material passes through the earthworm gut, a significant but not yet fully understood transformation takes place, whereby the resulting earthworm castings (worm manure) are rich in microbial activity and plant growth regulators, and fortified with pest repellence attributes as well! In short, earthworms, through a type of biological alchemy, are capable of transforming garbage into 'gold' (Vermi Co 2001, Tara Crescent 2003).

IMPORTANCE OF VERMICOMPOST

Source of plant nutrients

Earthworms consume various organic wastes and reduce the volume by 40–60%. Each earthworm weighs about 0.5 to 0.6 g, eats waste equivalent to its body weight and produces cast equivalent to about 50% of the waste it consumes in a day. These worm castings have been analyzed for chemical and biological properties. The moisture content of castings ranges between 32 and 66% and the pH is around 7.0. The worm castings contain higher percentage (nearly twofold) of both macro and micronutrients than the garden compost (Table 1).

Table 1. Nutrient composition of vermicompost and garden compost.

Nutrient element	Vermicompost (%)	Garden compost (%)
Organic carbon	9.8–13.4	12.2
Nitrogen	0.51–1.61	0.8
Phosphorus	0.19–1.02	0.35
Potassium	0.15–0.73	0.48
Calcium	1.18–7.61	2.27
Magnesium	0.093–0.568	0.57
Sodium	0.058–0.158	<0.01
Zinc	0.0042–0.110	0.0012
Copper	0.0026–0.0048	0.0017
Iron	0.2050–1.3313	1.1690
Manganese	0.0105–0.2038	0.0414

From earlier studies also it is evident that vermicompost provides all nutrients in readily available form and also enhances uptake of nutrients by plants. Sreenivas et al. (2000) studied the integrated effect of application of fertilizer and vermicompost on soil available nitrozen (N) and uptake of ridge gourd (Luffa acutangula) at Rajendranagar, Andhra Pradesh, India. Soil available N increased significantly with increasing levels of vermicompost and highest N uptake was obtained at 50% of the recommended fertilizer rate plus 10 t ha^{-1} vermicompost. Similarly, the uptake of N, phosphorus

(P), potassium (K) and magnesium (Mg) by rice (Oryza sativa) plant was highest when fertilizer was applied in combination with vermicompost (Jadhav et al. 1997).

Plant growth promoting activity

Growth promoting activity of vermicompost was tested using a plant bioassay method. The plumule length of maize (Zea mays) seedling was measured 48 h after soaking in vermicompost water and in normal water. The marked difference in plumule length of maize seedlings indicated that plant growth promoting hormones are present in vermicompost (Table 2).

Table 2. Plumule length of maize seedlings

Treatment	Initial length (cm)	Final length (cm)
Tank water	16.5	16.6
Vermicompost water	17.6	18.6

Improved crop growth and yield

Vermicompost plays a major role in improving growth and yield of different field crops, vegetables, flower and fruit crops. The application of vermicompost gave higher germination (93%) of mung bean (Vigna radiata) compared to the control (84%). Further, the growth and yield of mung bean was also significantly higher with vermicompost application. Likewise, in another pot experiment, the fresh and dry matter yields of cowpea (Vigna unguiculata) were higher when soil was amended with vermicompost than with biodigested slurry (Karmegam et al. 1999, Karmegam and Daniel 2000).

The efficiency of vermicompost was evaluated in a field study by Desai et al. (1999). They stated that the application of vermicompost along with fertilizer N gave higher dry matter (16.2 g plant^{-1}) and grain yield (3.6 t ha-1) of wheat (Triticum aestivum) and higher dry matter yield (0.66 g plant-1) of the following coriander (Coriandrum sativum) crop in sequential cropping system. Similarly, a positive response was obtained with the application of vermicompost to other field crops such as sorghum (Sorghum bicolor) (Patil and

Sheelavantar 2000) and sunflower (Helianthus annuus) (Devi and Agarwal 1998, Devi et al. 1998).

Application of vermicompost at 5 t ha^{-1} significantly increased yield of tomato (Lycopersicon esculentum) (5.8 t ha-1) in farmers' fields in Adarsha watershed, Kothapally, Andhra Pradesh compared to control (3.5 t ha^{-1}). Similarly, greenhouse studies at Ohio State University in Columbus, Ohio, USA have indicated that vermicompost enhances transplant growth rate of vegetables.

Amendment of vermicompost with a transplant grown without vermicompost had the highest amount of red marketable fruit at harvest. In addition, there were no symptoms of early blight lesions on the fruit at harvest. The yield of pea (Pisum sativum) was also higher with the application of vermicompost (10 t ha-1) along with recommended N, P and K than with these fertilizers alone (Reddy et al. 1998). Vadiraj et al. (1998) reported that application of vermicompost produced herbage yields of coriander cultivars that were comparable to those obtained with chemical fertilizers.

The fresh weight of flowers such as Chrysanthemum chinensis increased with the application of different levels of vermicompost. Also, the number of flowers per plant (26), flower diameter (6 cm) and yield (0.5 t ha-1) were maximum with the application of 10 t ha-1 of vermicompost along with 50% of recommended dose of NPK fertilizer. However, the vase life of flowers (11 days) was high with the combined application of vermicompost at 15 t ha-1 and 50% of recommended dose of NPK fertilizer (Nethra et al. 1999).

Reduction in soil C: N ratio

Vermicomposting converts household waste into compost within 30 days, reduces the C:N ratio and retains more N than the traditional methods of preparing composts (Gandhi et al. 1997). The C:N ratio of the unprocessed olive cake, vermicomposted olive cake and manure were 42, 29 and 11, respectively. Both the unprocessed olive cake and vermicomposted olive cake immobilized soil N throughout the study duration of 91 days. Cattle manure mineralized an appreciable amount of N during the study. The prolonged immobilization of soil N by the vermicomposted olive cake was attributed to the C:N ratio of 29 and to the recalcitrant nature of its C and N composition. The results suggest that for use of vermicomposted dry olive cake as an organic soil amendment, the management of vermicomposting process should be so adjusted as to ensure more favorable N mineralization immobilization (Thompson and Nogales 1999).

Role in nitrogen cycle

Earthworms play an important role in the recycling of N in different agroecosystems, especially under jhum (shifting cultivation) where the use of agrochemicals is minimal. Bhadauria and Ramakrishnan (1996) reported that during the fallow period intervening between two crops at the same site in 5- to 15-year jhum system, earthworms participated in N cycle through cast-egestion, mucus production and dead tissue decomposition. Soil N losses were more pronounced over a period of 15-year jhum system. The total soil N made available for plant uptake was higher than the total input of N to the soil through the addition of slashed vegetation, inorganic and organic manure, recycled crop residues and weeds.

Improved soil physical, chemical and biological properties

Limited studies on vermicompost indicate that it increases macropore space ranging from 50 to 500 μm, resulting in improved air-water relationship in the soil which favorably affect plant growth (Marinari et al. 2000). The application of organic matter including vermicompost favorably affects soil pH, microbial population and soil enzyme activities (Maheswarappa et al. 1999). It also reduces the proportion of water-soluble chemical species, which cause possible environmental contamination (Mitchell and Edwards 1997).

TYPES OF EARTHWORMS

Earthworms are invertebrates. There are nearly 3600 types of earthworms in the world and they are mainly divided into two types: (1) burrowing; and (2) non-burrowing. The burrowing types Pertima elongata and Pertima asiatica live deep in the soil. On the other hand, the non-burrowing types Eisenia fetida and Eudrilus eugenae live in the upper layer of soil surface. The burrowing types are pale, 20 to 30 cm long and live for 15 years. The non-burrowing types are red or purple and 10 to 15 cm long but their life span is only 28 months.

The non-burrowing earthworms eat 10% soil and 90% organic waste materials; these convert the organic waste into vermicompost faster than the burrowing earthworms. They can tolerate temperatures ranging from 0 to 40°C but the regeneration capacity is more at 25 to 30°C and 40–45% moisture level in the pile. The burrowing type of earthworms comes onto the soil surface only at night.

These make holes in the soil up to a depth of 3.5 m and produce 5.6 kg casts by ingesting 90% soil and 10% organic waste.

Earthworm multiplication

Numerous organic materials have been evaluated for growth and reproduction of earthworms as these materials directly affect the efficacy of vermicompost. Nogales et al. (1999) evaluated the suitability of dry olive cake, municipal biosolids and cattle manure as substrates for vermicomposting. They reported that larger weights of newly hatched earthworms were obtained in substrate containing dry olive cake. In another study, maize straw was found to be the most suitable feed material compared to soybean (Glycine max) straw, wheat straw, chickpea (Cicer arientinum) straw and city refuse for the tropical epigeic earthworm, Perionyx excavatus (Manna et al. 1997).

Zajonc and Sidor (1990) evaluated and compared various non-standard materials for the preparation of vermicompost. A mixture of cotton waste with cattle manure in the ratio of 1:5 was found to be the best. The use of grape cake alone increased earthworm weight slightly. Tobacco (Nicotiana tabacum) waste, used as substrate, increased earthworm weight but the earthworms failed to reproduce. A mixture of tobacco waste with rabbit manure in the ratio of 1:5 was found to be lethal to the earthworms.

A multiplication trial was conducted at the International Crops Research Institute for the Semi-Arid Tropics (ICRISAT), Patancheru, Andhra Pradesh with three kinds of earthworm cultures (Eisenia fetida, Eudrilus eugenae and Perionyx excavatus) using wheat straw, chickpea straw, tree leaves (Peltophorum sp) and Parthenium mixed with cow dung as feed materials. There was an increase in earthworm population and size during incubation for 90 days. The three types of earthworms multiplied 12 to 18 times when grown individually using legume tree leaves and cow dung mixture as raw material (Table 3). However, mixed culture (of all three species) showed higher multiplication rate (27 times) than the individual species.

Further studies on earthworm multiplication were also conducted at ICRISAT using tree leaves and Gliricidia stems mixed with cattle manure as feed material (Table 4). The earthworm population decreased when grown in mixture of Gliricidia stems and cattle manure. These results indicated that Gliricidia loppings could not be used for multiplication of earthworms. Gliricidia bark is known to possess toxic properties as it is used as rat poisoning bait.

In another multiplication study at ICRISAT, there was maximum increase in earthworm population (570%) and weight (109%) when grown in a feed material containing tree leaves (3 kg) and cow dung (6 kg). In contrast, mortality of earthworms (about 7 to 22%) was observed by growing them in a feed material containing soil (Table 5).

All these studies indicated that Gliricidia and tobacco leaves are not suitable for multiplication of earthworms. Perhaps the alkaloids and other principal compounds present in these leaves may effect the survival of earthworms. Also, soil and rabbit manure should not be mixed with earthworm feed material.

Table 3. Multiplication trial of earthworm species at ICRISAT, Patancheru, India in 2000[1]

Earthworm species	Initial population	Final population	Increase (%)
Mixed culture	900	15950	1612 (27)[2]
Eisenia fetida	90	1036	1051 (12)
Eudrilus eugenae	55	1007	1731 (18)
Perionyx excavatus	85	1192	1302 (14)

1. Mixture of legume tree leaves and cow dung was used as substrate.
2. Values in parentheses indicate increase in number of times at 90 days after incubation.

Table 4. Multiplication trials of earthworms using different organic materials at ICRISAT, Patancheru, India during 2000–02.

		Initial		Final[1]	
Earthworm species	Feed material	Population	Weight (g)	Population	Weight (g)
Eisenia fetida	Tree leaves (15 kg)	345	20	2510	207
	Cattle manure (15 kg)	510	207	1159	207
	Cattle manure (3 kg) + *Gliricidia* stem (6 kg)	1255	101	1000	50
Eudrilus eugenae	Tree leaves (15 kg)	311	21	2986	334
	Cattle manure (15 kg)	2986	334	1522	216
	Cattle manure (3 kg) + *Gliricidia* stem (6 kg)	2707	230	2249	100
Perionyx excavatus	Tree leaves (15 kg)	409	29	2707	230
	Cattle manure (15 kg)	2707	230	2650	187
	Cattle manure (3 kg) + *Gliricidia* stem (6 kg)	3356	365	1000	50

1. At 90 days after incubation.

Table 5. Multiplication trials of mixed culture of earthworms using soil and other organic substrates at ICRISAT, Patancheru, India, 2000–02

	Initial		Final		Increase[1] (%)	
Feed material	Number	Weight (g)	Number	Weight (g)	Number	Weight
Cow dung (15 kg)	500	89	750	163	50	83
Tree leaves (3 kg) + cow dung (3 kg)	500	95	1545	125	21	32
Tree leaves (3 kg) + cow dung (6 kg)	500	110	3351	230	570	109
Pigeonpea leaves + pod shells + tree leaves (2 kg) + cow dung (2 kg)	500	98	2230	187	346	90
Pigeonpea leaves + pod shells + tree leaves (2 kg) + cow dung (4 kg)	500	115	1490	193	198	68
Soil (5 kg) + cow dung (5 kg)	1000	90	784	87	–22	–3
Soil (5 kg) + cow dung (5 kg) + pigeonpea leaves (1 kg)	1000	75	1023	241	2	223
Soil (5 kg) + cow dung (5 kg) + tree leaves (1 kg)	1000	160	929	170	–7	–6

1. At 90 days after incubation

TEMPERATURE CHANGES DURING THE PROCESS

Change in temperature was observed during the process of vermicomposting (from 5 to 65 days) with different farm residues (Parthenium and grass). In the beginning of the process, ie, up to 15 days, the temperature was high (32 to 33°C) in both Parthenium and grass substrates when compared to outside temperature (26 to 30°C). Later, there was a gradual decrease in temperature, which reached a minimum of about 24°C. However, higher temperature was recorded in Parthenium compost (decline from 32.8 to 27.5°C) than in grass compost (decline from 31.5 to 26.8°C) during the whole period of digestion process. Generally more heat was evolved from control treatment (without earthworms) than the vermicompost treatments (with earthworms). From these studies, it was suggested that the most suitable period for releasing the earthworms into organic residues would be between 15 and 20 days after heaping of the organic residues when the temperature is about 25°C (Fig. 1).

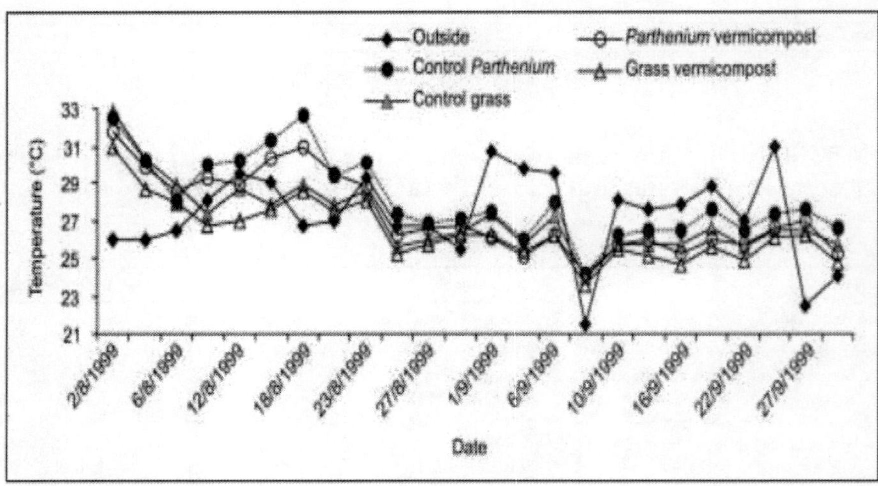

Figure 1. Temperature changes during biodigestion.

METHODS OF VERMICOMPOSTING

Pits below the ground

Pits made for vermicomposting are 1 m deep and 1.5 m wide. The length varies as required.

Heaping above the ground

The waste material is spread on a polythene sheet placed on the ground and then covered with cattle dung. Sunitha et al. (1997) compared the efficacy of pit and heap methods of preparing vermicompost under field conditions. Considering the biodegradation of wastes as the criterion, the heap method of preparing vermicompost was better than the pit method. Earthworm population was high in the heap method, with a 21-fold increase in Eudrilus eugenae as compared to 17-fold increase in the pit method. Biomass production was also higher in the heap method (46-fold increase) than in the pit method (31-fold). Consequent production of vermicompost was also higher in the heap method (51 kg) than in the pit method (40 kg).

Tanks above the ground

Tanks made up of different materials such as normal bricks, hollow bricks, shabaz stones, asbestos sheets and locally available rocks were evaluated for vermicompost preparation. Tanks can be constructed with the dimensions suitable for operations. At ICRISAT, we have evaluated tanks with dimensions of 1.5 m (5 feet) width, 4.5 m (15 feet) length and 0.9 m (3 feet) height. The commercial biodigester contains a partition wall with small holes to facilitate easy movement of earthworms from one tank to the other.

Cement rings

Vermicompost can also be prepared above the ground by using cement rings (ICRISAT and APRLP 2003). The size of the cement ring should be 90 cm in diameter and 30 cm in height. The details of preparing vermicompost by this method have been described in a later section.

Commercial model

The commercial model for vermicomposting developed by ICRISAT consists of four chambers enclosed by a wall (1.5 m width, 4.5 m length and 0.9 m height) (Fig. 2). The walls are made up of different materials such as normal bricks, hollow bricks, shabaz stones, asbestos sheets and locally available rocks. This model contains partition walls with small holes to facilitate easy movement of earthworms from one chamber to another. Providing an outlet at one corner of each chamber with a slight slope facilitates collection of excess water, which is reused later or used as earthworm leachate on crop. The outline of the commercial model is given in Figure 3.

The four components of a tank are filled with plant residues one after another. The first chamber is filled layer by layer along with cow dung and then earthworms are released. Then the second chamber is filled layer by layer. Once the contents in the first chamber are processed the earthworms move to chamber 2, which is already filled and ready for earthworms. This facilitates harvesting of decomposed material from the first chamber and also saves labor for harvesting and introducing earthworms. This technology reduces labor cost and saves water as well as time.

Figure 2. Commercial model for vermicomposting at ICRISAT

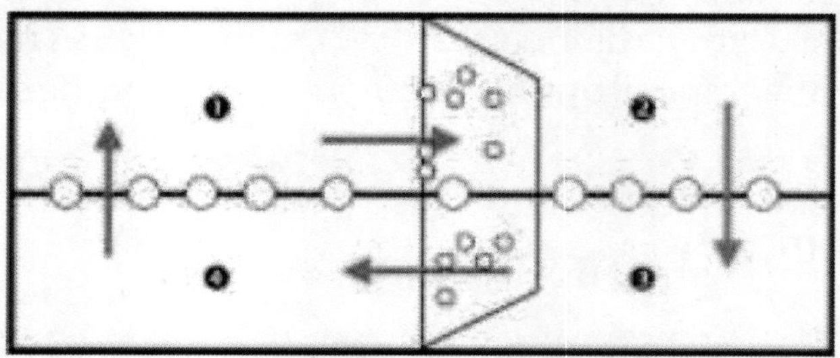

Figure 3. Diagrammatic representation of the commercial model with four chambers for vermicomposting.

Materials Required for Vermicomposting

A range of agricultural residues, all dry wastes, for example, sorghum straw and rice straw (after feeding cattle), dry leaves of crops and trees, pigeonpea (Cajanus cajan) stalks, groundnut (Arachis hypogaea) husk, soybean residues, vegetable wastes, weed (Parthenium) plants before flowering, fiber from coconut (Cocos nucifera) trees and sugarcane (Saccharum officinarum) trash can be converted into vermicompost. In addition, animal manures, dairy and poultry wastes, food industry wastes, municipal solid wastes, biogas sludge and bagasse from sugarcane factories also serve as good raw materials for vermicomposting.

The quantity of raw materials required using a cement ring of 90 cm in diameter and 30 cm in height or a pit or tank measuring 1.5 m × 1 m × 1 m is given below:

Dry organic wastes (DOW)	50 kg
Dung slurry (DS)	15 kg
Rock phosphate (RP)	2 kg
Earthworms (EW)	500–700
Water (W)	5 L every three days

The various ingredients are used in the ratio of 5:1.5:0.2:50–75:0.5 of DOW:DS:RP: EW:W. In the tank or pit system 100 kg of raw material and 15–20 kg of cow dung are needed for each cubic meter of the bed

VERMICOMPOST PREPARATION

Steps in the process

Vermicomposting involves the following steps which are depicted in Figure 4(a–k):

- Cover the bottom of the cement ring with a layer of tiles or coconut husk or polythene sheet (Fig. 4a).
- Spread 15–20 cm layer of organic waste material on the polythene sheet (Fig. 4b). Sprinkle rock phosphate powder if available (it helps in improving nutritional quality of compost)

on the waste material and then sprinkle cow dung slurry (Fig. 4c and d). Fill the ring completely in layers as described. Paste the top of the ring with soil or cow dung (Fig. 4e). Allow the material to decompose for 15 to 20 days.

- When the heat evolved during the decomposition of the materials has subsided (15–20 days after heaping), release selected earthworms (500 to 700) through the cracks developed (Fig. 4f).
- Cover the ring with wire mesh or gunny bag to prevent birds from picking the earthworms.

Sprinkle water every three days to maintain adequate moisture and body temperature of the earthworms (Fig. 4g).

- The vermicompost is ready in about 2 months if agricultural waste is used and about 4 weeks if sericulture waste is used as substrate (Fig. 4h).
- The processed vermicompost is black, light in weight and free from bad odor.
- When the compost is ready, do not water for 2–3 days to make compost easy for sifting. Pile the compost in small heaps and leave under ambient conditions for a couple of hours when all the worms move down the heap in the bed (Fig. 4i). Separate upper portion of the manure and sieve the lower portion to separate the earthworms from the manure (Fig. 4j). The culture in the bed contains different stages of the earthworm's life cycle, namely, cocoons, juveniles and adults. Transfer this culture to fresh half decomposed feed material. The excess as well as big earthworms can be used for feeding fish or poultry. Pack the compost in bags and store the bags in a cool place (Fig. 4k).
- Prepare another pile about 20 days before removing the compost and repeat the process by following the same procedure as described above.

Precautions during the process

The following precautions should be taken during vermicomposting:

- The African species of earthworms, Eisenia fetida and Eudrilus eugenae are ideal for the preparation of vermicompost. Most

Indian species are not suitable for the purpose.

- Only plant-based materials such as grass, leaves or vegetable peelings should be utilized in preparing vermicompost.
- Materials of animal origin such as eggshells, meat, bone, chicken droppings, etc are not suitable for preparing vermicompost.
- Gliricidia loppings and tobacco leaves are not suitable for rearing earthworms.
- The earthworms should be protected against birds, termites, ants and rats.
- Adequate moisture should be maintained during the process. Either stagnant water or lack of moisture could kill the earthworms.
- After completion of the process, the vermicompost should be removed from the bed at regular intervals and replaced by fresh waste materials.

How to Use Vermicompost?

- Vermicompost can be used for all crops: agricultural, horticultural, ornamental and vegetables at any stage of the crop.
- For general field crops: Around 2–3 t ha-1 vermicompost is used by mixing with seed at the time of sowing or by row application when the seedlings are 12–15 cm in height. Normal irrigation is followed.
- For fruit trees: The amount of vermicompost ranges from 5 to 10 kg per tree depending on the age of the plant. For efficient application, a ring (15–18 cm deep) is made around the plant. A thin layer of dry cow dung and bone meal is spread along with 2–5 kg of vermicompost and water is sprayed on the surface after covering with soil.
- For vegetables: For raising seedlings to be transplanted, vermicompost at 1 t ha-1 is applied in the nursery bed. This results in healthy and vigorous seedlings. But for transplants, vermicompost at the rate of 400–500 g per plant is applied initially at the time of planting and 45 days after planting (before irrigation).

- For flowers: Vermicompost is applied at 750–1000 kg ha-1
- For vegetable and flower crops vermicompost is applied around the base of the plant. It is then covered with soil and watered regularly

Figure 4(a–k). Vermicomposting process.

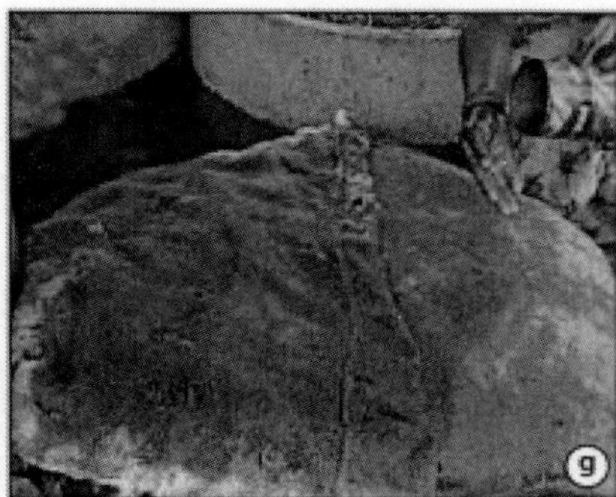

Cement ring covered with gunny bag

Processed vermicompost

Heaping of vermicompost

Compost sieved

Bag filled with vermicompost

BIODIVERSITY IN VERMICOMPOST

In the present study, vermicompost samples were collected and analyzed for microbial diversity and population studies. The vermicompost samples were collected in sterile containers from the rings before harvesting the compost. To compare microbial diversity, samples from the partially decomposed dry organic waste material, ready for the release of the earthworms, were also collected and checked for diversity and population counts.

Total mircobial populations of bacteria, fungi and actinomycetes from the substrates were determined by using dilution plate techniques with suitable media (Nutrient Agar, Potato Dextrose Agar, Actinomycetes Isolation Agar-HI Media). The number of colony forming units (CFU) was expressed as CFU g^{-1}

Several authors have noted that the earthworms play a major role in affecting populations of soil organisms, especially in causing changes in the soil microbial community (Coleman 1985, Parmelee 1998). The present work recorded higher microbial populations in the partially decomposed dry organic waste material for vermicompost than the vermicompost (Table 6). This may be due to the existing temperatures and pH in the partially decomposed raw material. But compared to conventional thermophilic composts, vermicompost is much richer in microbial diversity, populations and activities (Subler et al. 1998).

Table 6. Microbial populations from the samples of vermicompost.

	Bacteria (CFU g^{-1})	Fungi (CFU g^{-1})	Actinomycetes (CFU g^{-1})
Vermicompost	54×10^6	8×10^4	1×10^4
Partially decomposed dry organic waste material for vermicompost	69×10^6	11×10^4	2×10^4

The fungal isolates from the samples were identified upto species level (Table 7). Much diversity was observed between the two samples collected. Aspergillus, Fusarium, Mucor, Cladosporium, and Trichoderma were the common genera observed in both the samples. Genera like Absidia, and Stachbotrys were recorded in vermicompost. Genera like Alternaria, penicillium, and Thermomyces were isolated from partially decomposed dry organic waste material for vermicompost. This clearly indicates that the fungal diversity is more in the decomposed material than in the vermicompost. The digestive epithelium of the simple straight tubular gut of worms is known to secrete cellulase, amylase, invertase, protease, phosphatase (Ranganathan and Vinotha 1998). Earthworms inevitably consume the soil microbes during the ingestion of litter and soil. It has been recently estimated that earthworms necessarily have to feed on microbes, particularly fungi for their protein/nitrogen requirement (Ranganathan and Parthasarathi 2000). This may be the reason for the less diversity of fungi and microbial counts seen in the vermicompost collected

In both the samples percentage of Aspergillus was more when compared with other genera. Tricoderma and Penicillium have antibiotic activities and can also be used as biological control on soil borne pathogens. Only a few studies have investigated that the suppression of soil borne plant pathogens by vermicompost (Szczech et al. 1993), or disease suppression in the presence of earthworms (Stephens and Davoren 1997, Stephens et al. 1994). Disease suppression by compost has been attributed to the activities of competitive or antagonistic microorganisms as well as the antibiotic compounds present in the vermicompost.

Table 7. List of fungi isolated from partially decomposed dry organic waste for vermicompost and vermicompost.

artially decomposed dry organic waste r vermicompost	Vermicompost
lternaria citri	*Absidia cylindrospora*
spergillus fumigatus	*Aspergillus fumigatus*
spergillus niger	*Aspergillus niger*
spergillus cervinus	*Aspergillus clavoto nanicus*
spergillus terreus	*Aspergillus terreus*
spergillus sydowii	*Aspergillus sydowii*
spergillus niveus	*Aspergillus nidulans*
spergillus sclerotiorum	*Cladosporium herbarum*
ladosporium cladosporioides	*Fusarium oxysporum*
ladosporium herbarum	*Fusarium semitactum*
usarium samucinum	*Fusarium nivale*
usarium dimerum	*Mucor circinelloides*
lucor racemosus	*Stachbotrys chartarum*
nicillium chrysogenum	*Trichoderma viride*
nicillium thomii	
nicillium citrinum	
ichoderma viride	
hermomyces lanuginous	

Vermicomposting: A Livelihood Micro-enterprise for Rural Women

ICRISAT with support from the Asian Development Bank (ADB), Philippines, District Water Management Agency (DWMA), Government of Andhra Pradesh and Tata-ICRISAT-ICAR project in northeastern regions of India was keen to promote the vermiculture technology. The primary objective of this project was to help women from rural areas to set up micro-enterprises based on vermiculture technology and also to improve crop productivity by increasing soil fertility through ecological methods of farming (Wani 2002).

The training program conducted by ICRISAT for DWACRA (Development of Women and Child in Rural Area) group of women and other women self-help groups (SHGs) covered technical aspects of multiplying earthworms, managing and collection of organic wastes, application of vermicompost for various crops, accounting and marketing. At the same time a noxious weed, Parthenium hysterophorus (locally referred as vayyari bhama or congress weed), was found abundantly in the fields as well as on field bunds, which inhibited crop growth and caused environmental pollution. Hence, the women have come forward to utilize this weed as raw material for vermicomposting, which is a safe weed disposal mechanism and an opportunity to convert into valuable compost.

CASE STUDIES

Adarsha watershed, Kothapally

Ms Lakshmamma and four other women have set up a vermicomposting enterprise in a common place under one roof. Having begun with a population of 2,000 earthworms of three epigeic species, they regularly harvest around 400 kg of vermicompost every month collectively. Their work in making vermicompost is shared collectively and the unique marketing strategy involves meeting potential customers. Sometimes, they even get customers from distant places. They earn a net income of around Rs 500 each month. By becoming an earning member of the family, they are involved in the decision-making process in the family. This has also raised their status in the society

APRLP watershed village

Ms Padmamma living in Sripuram, one of the thousand non-descript villages of Mahbubnagar district in Andhra Pradesh, leads a routine life and has never dreamt of a different life. She joined the women's SHG at the begining of the Andhra Pradesh Rural Livelihoods Programme (APRLP) project. Though reluctant during the initial stage, she started taking active part in the weekly meetings and showed interest in the discussions about raising income through small initiatives like adopting the vermicompost scheme. This scheme was introduced to enhance crop productivity in the fields and enable the farmers to get more per-hectare yield. Ms Padmamma is able to get higher yield from different crops such as maize and vegetables with the application of vermicompost in her own field. She now proudly displays the vermiculture beds to any visitor who comes to meet her.

Tata-ICRISAT-ICAR project

The farmers of Bundi nucleus watershed in Rajasthan, India have shown lot of interest in vermicomposting. Two farmers have built a multiple compartment system (commercial model) of vermicomposting while many are following the regular vermicomposting. In Guna nucleus watershed in Madhya Pradesh, nearly 35 farmers from all the three microwatersheds are practicing vermicomposting.

Most of them are producing vermicompost on a large scale and are applying to their own fields for vegetable crops and getting higher yields with low-cost technology. A few farmers have already started selling their extra produce of vermicompost at the nearby market at the rate of Rs 5–7 per kg.

CONCLUSIONS

The production of degradable organic waste and its safe disposal becomes the current global problem. Meanwhile the rejuvenation of degraded soils by protecting topsoil and sustainability of productive soils is a major concern at the international level. Provision of a sustainable environment in the soil by amending with good quality organic soil additives enhances the water holding capacity and nutrient supplying capacity of soil and also the development of resistance in plants to pests and diseases. By reducing the time of humification process and by evolving the methods to minimize the loss of nutrients during the course of decomposition, the fantasy becomes fact. Earthworms can serve as tools to facilitate these functions. They serve as "nature's plowman" and form nature's gift to produce good humus, which is the most precious material to fulfill the nutritional needs of crops. The utilization of vermicompost results in several benefits to farmers, industries, environment and overall national economy.

To farmers

- Less reliance on purchased inputs of nutrients leading to lower cost of production
- Increased soil productivity through improved soil quality
- Better quantity and quality of crops
- For landless people provides additional source of income generation

To industries:

- Cost-effective pollution abatement technology

To environment:

- Wastes create no pollution, as they become valuable raw materials for enhancing soil fertility

To national economy

- Boost to rural economy
- Savings in purchased inputs
- Less wasteland formation

REFERENCES

1. Aveyard Jim. 1988. Land degradation: Changing attitudes - why? Journal of Soil Conservation, New South Wales 44:46–51.
2. Bhadauria T and Ramakrishnan PS. 1996. Role of earthworms in nitrogen

cycle during the cropping phase of shifting agriculture (jhum) in northeast India. Biology and Fertility of Soils 22:350–354.

3. Bhiday MR. 1994. Earthworms in agriculture. Indian Farming 43(12):31–34.

4. Coleman D C. 1985. Through a red darkly: an ecological assessment of root soil microbial faunal interactions. Pages 1–21 in Ecological interaction in Soil (Fitter AH, Atkinson D, Read DJ and Usher MB, eds.). London, UK: Blackwell Scientific Publications.

5. Desai VR, Sabale RN and Raundal PV. 1999. Integrated nitrogen management in wheat-coriander cropping system. Journal of Maharasthra Agricultural Universities 24(3):273–275.

6. Devi D and Agarwal SK. 1998. Performance of sunflower hybrids as influenced by organic manure and fertilizer. Journal of Oilseeds Research 15(2):272–279.

7. Devi D, Agarwal SK and Dayal D. 1998. Response of sunflower [Helianthus annuus (L.)] to organic manures and fertilizers. Indian Journal of Agronomy 43(3):469–473.

8. Gandhi M, Sangwan V, Kapoor KK and Dilbaghi N. 1997. Composting of household wastes with and without earthworms. Environment and Ecology 15(2):432–434.

9. ICRISAT and APRLP. 2003. Vermicomposting: Conversion of organic wastes into valuable manure. Andhra Pradesh, India: ICRISAT and APRLP. 4 pp.

10. Jadhav AD, Talashilkar SC and Pawar AG. 1997. Influence of the conjunctive use of FYM, vermicompost and urea on growth and nutrient uptake in rice. Journal of Maharashtra Agricultural Universities 22(2):249–250.

11. Karmegam N, Alagermalai K and Daniel T. 1999. Effect of vermicompost on the growth and yield of greengram (Phaseolus aureus Rob.). Tropical Agriculture 76(2):143–146.

12. Karmegam N and Daniel T. 2000. Effect of biodigested slurry and vermicompost on the growth and yield of cowpea [Vigna unguiculata (L.)]. Environment and Ecology 18(2):367–370.

13. Maheswarappa HP, Nanjappa HV and Hegde MR. 1999. Influence of organic manures on yield of arrowroot, soil physico-chemical and biological properties when grown as intercrop in coconut garden. Annals of Agricultural Research 20(3):318–323.

14. Manna MC, Singh M, Kundu S, Tripathi AK and Takkar PN. 1997. Growth and reproduction of the vermicomposting earthworm Perionyx excavatus as influenced by food materials. Biology and Fertility of Soils 24(1):129–132.

15. Marinari S, Masciandaro G, Ceccanti B and Grego S. 2000. Influence of organic and mineral fertilisers on soil biological and physical properties. Bioresource Technology 72(1):9–17.

16. Mitchell A and Edwards CA. 1997. The production of vermicompost using Eisenia fetida from cattle manure. Soil Biology and Biochemistry 29:3–4.

17. Nethra NN, Jayaprasad KV and Kale RD. 1999. China aster [Callistephus

chinensis (L)] cultivation using vermicompost as organic amendment. Crop Research, Hisar 17(2): 209–215.

18. Nogales R, Melgar R, Guerrero A, Lozada G, Beniteze E, Thompson R, Gomez M and Garvin MH. 1999.

19. Growth and reproduction of Eisenia andrei in dry olive cake mixed with other organic wastes. Pedobiologia 43(6):744–752.

20. Parmelee RW, Bohlen PJ and Blair JM. 1998. Earthworms and nutrient cycling processes: intergrating across the ecological hierarchy. Pages 123–143 in Earthworm Ecology (Edwards CA, ed.). New York, USA: St Lucie Press.

21. Patil SL and Sheelavantar MN. 2000. Effect of moisture conservation practices, organic sources and nitrogen levels on yield, water use and root development of rabi sorghum [Sorghum bicolor (L.)] in the vertisols of sem arid tropics. Annals of Agricultural Research 21(21):32–36.

22. Ranganathan LS and Parthasarathi K. 2000. Enhanced phosphatase activity in earthworm casts is more of microbial origin. Current Science 79: 1158–1159.

23. Ranganathan LS and Vinotha SP. 1998. Influence of pressmud on the enzymatic variations in the different reproductive stages of Eudrilus eugeniae. Current Science 74: 634–635.

24. Reddy R, Reddy MAN, Reddy YTN, Reddy NS, Anjanappa N and Reddy R. 1998. Effect of organic and inorganic sources of NPK on growth and yield of pea [Pisum sativum(L)]. Legume Research 21(1):57–60.

25. Sreenivas C, Muralidhar S and Rao MS. 2000. Vermicompost, a viable component of IPNSS in nitrogen nutrition of ridge gourd. Annals of Agricultural Research 21(1):108–113.

26. Sunitha ND, Giraddi RS, Kulkarni KA and Lingappa S. 1997. Evaluation methods of vermicomposting under open field conditions. Karnataka Journal of Agricultural Sciences 10(4): 987–990.

27. Stephens PM, Davoren CW, Ryder MH, Doube BM and Correll RL. 1994. Field evidence for reduced severity of Rhizoctonia bare-patch disease of wheat, due to the presence of the earthworms Aporrectodea rosea and Aporrectodea trapezoides. Soil Biology and Biochemistry 26(11): 1495–1500.

28. Subler S, Edwards C A and Metzger J. 1998. Comparing vermicomposts and composts. BioCycle 39: 63–66.

29. Szczech M, Rondomanski W, Brzeski MW, Smolinska U and Kotowski J. 1993. Suppressive effect of commercial earthworm compost on some root infecting pathogens of cabbage and tomato. Biological Agriculture and Horticulture 10(1): 47–52.

30. Tara Crescent. 2003. Vermicomposting. Development Alternatives (DA) Sustainable Livelihoods. (http://www.dainet.org/livelihoods/default.htm)

31. Thompson RB and Nogales R. 1999. Nitrogen and carbon mineralization in soil of vermicomposted and unprocessed dry olive cake ('Orujo seco') produced from two stage centrifugation for olive oil extraction.

32. Journal of Environmental Science and Health, Part B, Pesticides, Food Contaminants and Agricultural Wastes 34(5):917–928.

33. Vadiraj BA, Siddagangaiah D and Potty SN. 1998. Response of coriander (Coriandrum sativum L) cultivars to graded levels of vermicompost. Journal of Spices and Aromatic Crops 7(2):141–143.

34. Vermi Co. 2001. Vermicomposting technology for waste management and agriculture: an executive summary. (http://www.vermico.com/summary.htm) PO Box 2334, Grants Pass, OR 97528, USA: Vermi Co.

35. Wani SP. 2002. Improving the livelihoods: New partnerships for win-win solutions for natural resource management. Paper submitted in the 2nd International Agronomy Congress held at New Delhi, India during 26 30 November 2002.

36. Wani SP and Lee KK. 1992. Biofertilizers role in upland crops production. Pages 91–112 in Fertilizers, organic manures, recyclable wastes and biofertilisers (Tandon HLS, ed.). New Delhi, India: Fertilizer Development and Consultation Organisation.

37. Wani SP, Rupela OP and Lee KK. 1995. Sustainable agriculture in the semi-arid tropics through biological nitrogen fixation in grain legumes. Plant and Soil 174:29–49.

38. Zajonc I and Sidor V. 1990. Use of some wastes for vermicompost preparation and their influence on growth and reproduction of the earthworm Eisenia fetida. Pol'nohospodars-tvo (CSFR) 36(8):742–752.

Chapter 13

LONG-TERM FIELD EXPERIMENT IN SWEDEN: EFFECTS OF ORGANIC AND INORGANIC FERTILIZERS ON SOIL FERTILITY AND CROP QUALITY

Artur Granstedt[1] & Lars Kjellenberg[2]

Tufts University, Agricultural Production and Nutrition, Massachusetts

In 1958, Bo D. Pettersson in the Nordic Research Circle for Biodynamic Farming in Järna, Sweden, began an agricultural field experiment that lasted until 1990, i.e. 32 years. The field experiment included eight different fertilizer treatments, each with a four-year crop rotation without repetitions: summer wheat, clover/grass mix, potatoes, beets. The focus was primarily on aspects of crop quality, and the fertilizer application rates for the various treatments were adjusted to bring about comparable yields. Two "daughter experiments" emerged from the K-experiment and were run in parallel with the mother project during 1971-1976 in Uppsala and 1971-1979 in Järna.

In these experiments a comparison was made between two systems, biodynamic farming and conventional farming, in which both fertilizer regimes and crop rotations were studied. One of the main objectives in the K-experiment, i.e. to obtain nearly the same yield over the experimental period in the organic-treatment variants and in the inorganic treatments has largely been achieved, but there were differences between crops.

During the time between 1958 and 1990 the yield increased in all treatments in accordance with the overall trend in the Swedish agriculture, but the increase was highest in the organic treatments (65 % in the biodynamic in comparison with 50 % in the conventional). The effects of the different fertilizer treatments on product quality are in accordance with findings in the two "daughter experiments" which were based on the original K-experiment.

Compared with the conventional treatments, the crude protein content of potatoes and wheat was lower in the organic treatments, but protein quality was higher (i.e. relatively pure protein and essential amino acids, lower amount of free amino acids). Resistance to decomposition and store quality for potatoes were higher in the organic treatments, and in wheat starch quality seemed to be higher.

The organic treatments resulted in a higher soil fertility capacity and in crops with higher quality protein, a higher starch content, and a greater ability to tolerate stressful conditions and long-term storage in comparison with the inorganic treatments. Furthermore, the crops produced in the organic treatments developed a structure that can be studied through a picture formation method (Crystallization with CuCl2). This has also been described as a higher organizational level which is evident in terms of both soil and crop formation as a result of the long-term effects of organic manure compared with conventional NPK-fertilizer.

New experiments in Sweden and Finland have been started to study the effects of different organic treatments on farms. Preliminary results of these experiments confirm the described differences between organic and inorganic treatments, but indicate also that the effects of liquid organic manure on quality parameters are more similar to those of inorganic fertilizer[1] Agricultural Research Centre of Finland, Partala Research Station for Ecological Agriculture, FIN-51900 Juva, Finland[2]Biodynamic Research Institute, Skilleby, S-153 00 Järna, Sweden

INTRODUCTION

In 1958, Bo D Pettersson and the Scandinavian Research Circle in Järna, Sweden, began an agricultural field experiment to determine in what ways various types of fertilizers affect the soil and final quality of grain and vegetable products. Various quality parameters and quality-assessment methods were developed and tested during the experiment period which spanned 32 years (up to 1990), and several reports have been published (Pettersson, Reents & Wistinghausen, 1992). Two "daughter experiments" emerged from the K-experiment and were run in parallel with the mother project during 1971-1976 in Uppsala and 1971-1979 in Järna (referred to as UJ-experiments in the following text). In these experiments a comparison was made between two systems, biodynamic farming and conventional farming, in which both fertilizer regimes and crop rotations were studied (Pettersson, 1982; Dlouhy, 1981). Before that, the influence of these systems on quality parameters for potatoes under different climatic and soil conditions also had been studied in different parts of Scandinavia (Pettersson, 1970).

Description of the K-experiment

The K experiment was located at 59° North, 17° East at an elevation of 10 m above sea level. The mean yearly precipitation was 550 mm, and the mean annual temperature was 6 degrees C, with 6-8 snow-free months per year. The soil was a silty loam with an intermediate humus content. Experimental Layout: To ensure that the field experiment could be used for the plant quality assessments while providing the flexibility to support other experiments that had yet to to designed, a very broad basis was adopted. This scheme included eight different fertilizer treatments, each with a 4-fold crop rotation without repetitions. The size of each subplot was 36 m^2 gross, with a net harvestable area of 27 m^2.The Crop Rotation: Within each fertilizer variant the following crops were rotated without interruption so that in any given year all four would be present: summer wheat (undersown with clover/grass), clover/grass mix, potatoes and beets. Fertilization Scheme: To facilitate focusing primarily on aspects of crop quality, the fertilizer application rates for the various treatments were adjusted to bring about comparable yields. This applies to the variants 1,2,3,4,7 and 8; variant 5 was not fertilized at all (Tables 1 and 2): Table 1. The fertilization program (application rates of N- tot, P and K in kg/ha/yr averaged for the years 1958-1990 in parentheses).

K1.	Composted manure (82/38/76).: aged half a year with the addition of biodynamic compost preparations 502-507 and biodynamic preparation sprays 500 and 501.
K2.	Composted manure (82/38/76): same as in K1 with the exception that treatments with biodynamic preparation sprays 500 and 501 were excluded.
K3.	Raw manure (95/30/91): with horn and bone meal added to reach 1% contents.
K4.	Raw manure + NPK (63/28/66): half the K3 manure rate; half the K6 NPK rate, resp.
K5.	Control: unfertilized.
K6.	Inorganic NPK (19/19/41)
K7.	Inorganic NPK(59/36/81): twice the rate applied in K6.
K8.	Inorganic NPK(114/36/81): as in K6 but 4 times the N rate and twice the P and K rates.

Table 2. Breakdown of the fertilization scheme within the rotation, in %.

Variant	Wheat	Type	Clover/grass	Potatoes	Beets
K1, 2 & 3	Organic	-	-	40	60
K6, 7 & 8	P K	-	-	40	60
	N	20	-	40	40

Description of the UJ-experiment

In the UJ experiment conventional (A) and biodynamic (B) treatments were compared with each other in two crop rotations as described below for the experiment in Järna (table 3): Crop rotation 1 represented a rotation without animals, and crop rotation 2 represented a system with animals related to an organic farming system which under

Nordic conditions is self-sufficient with fodder with 0.8 CU (cattle units)/ha (average 50 kg N/ha and year). The field conditions were almost the same as those in the K-experiment. A split-split-plot design was used with three replications. All crops were grown each year. The soil was a silty loam well supplied with plant nutrients but with a low humus content. The weather was drier than normal for the region during the first years of the experiment period and had a strongly negative effect on the yields of the grain and ley.

During 1971-1976 a parallel project was carried out in Uppsala (called UJ-experiment Ultuna) with the same treatments as in Järna (for more details see Dlouhy, 1981). The soil was an intermediate clay, well supplied with plant nutrients and moderately rich in humus.

Table 3. Crop rotation and fertilization scheme in UJ Experiment in Järna 1971-1979.

CONVENTIONAL SYSTEM ARTIFICIAL FERTILIZERS HERBICIDES AND PESTICIDES				BIODYNAMIC SYSTEM ORGANIC MANURE			
Crop rotation A1	Nutrient application			Crop rotation B1	Nutrient application		
	kg	ha	yr		kg	ha	yr
	N	P	K		N	P	K
Barley	80	20	35	Barley	60	50	55
Potatoes	120	100	265	Potatoes	100	65	95
Spring wheat	80	20	35	Spring wheat	50	30	45
Crop rotation A2				Crop rotation B2			
Ley				Ley			
Potatoes	100	80	225	Potatoes	120	80	110
Spring wheat	40	30	50	Spring wheat	70	60	65

Yield and soil fertility

'Yield in the K-experiment. The summer wheat and beet yields were increased by fertilization, and a considerable difference was also found between the K1 (compost + B-D sprays) and K2 (compost without sprays) variants. Yields of clover/grass declined in response to fertilizer treatment and were highest in the organic variants. A decrease in the yield of the legume mix can be expected to result in a reduction in nitrogen fixation (Table 4). In the K5 variant that did not receive any fertilizer for 30 years, the yield of ley was on the same level as that in the other treatments, and also here, the yield tended to increase over the period, although the increase was not as pronounced. Table 4. Average yields during 1958-1989 in dt/ha/yr for single crops

Variant	S. wheat	Clover/grass	Potatoes	Beets	Beet leaves
	(15% moist.)	1st cut	(80 % moist.)		(85 % moist)
K1.	32.7	48.9	362	467	363
K2.	29.8	50.2	355	451	362
K3.	32.6	51.5	352	475	355
K4.	32.6	50.6	365	450	365
K5.	24.9	42.7	287	213	192
K6.	30.2	44.8	343	363	295
K7.	33.1	42.6	370	456	344
K8.	32.8	43.0	362	493	454
Mean.	31.1	46.6	349	421	341

In terms of their 32-year averages, the yields are all comparable with the exception of the control variant (K5) and excluding data from the conversion period for the organic treatments. All yields except for those of the unfertilized variant (K5) increased with successive rotation periods (figure 1). The figures indicate that there was a conversion period of about 8 years during which the yield level was lower in the organic treatments. A

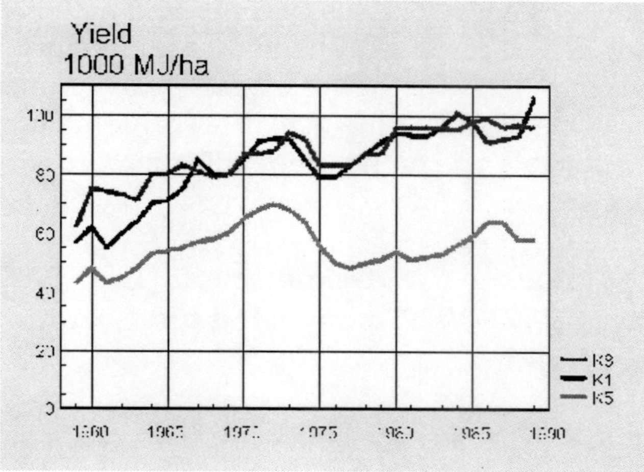

Figure 1. K-experiment 1958-1989. Yield, MJ/haDevelopment of soil fertility.

Data on soil physical properties, soil chemistry and soil biology were collected after 19, 28 and 32 years in the K-experiment (Pettersson, Reents and Wistinghausen, 1992). The analysis of soil data shows that nearly all chemical (pH, P, K, Mg, C and N) (macronutrient availability was an exception) and biological parameters (respiration, DHA, urease, earthworms) assessed were improved by organic fertilization whereas no such improvements were observed following the application of mineral fertilizer. Similar results concerning soil parameters were obtained in the UJ-experiment (Pettersson, 1982). In the UJ-experiment at Järna, humus contents had increased significantly after 9 years (by more than 10 percent) in B2 compared with the recent value of 2.72 percent and the value of the conventional treatment of 2.74 percent (A2B2 $P<0.001$).

Mineralization capacity.

The mineralization capacity of the soil was studied in summer wheat after potatoes (the third year after ley) in the K-experiment. Net mineralization in the soil was estimated by measuring the mineral nitrogen content of soil samples taken in spring and at harvest and by determining the total nitrogen uptake of the spring wheat crop (Granstedt, 1992). The mean mineralization capacity during 1988-1990 was, on average, 95% higher (i.e. 106%, 146% and 33% higher) in the organically farmed treatments as compared with the treatments receiving commercial fertilizer. The higher mineralization capacity compensated completely for the absence of applied mineral nitrogen in the organically farmed treatment. In the organic manure system we built up a more stable and higher mineralization capacity compared with the commercially fertilized treatments.

Quality of potatoes in the K-experiment (1958-1989) and UJ-experiment (1971-1979) comparing biodynamic and conventional farming.

The main aims of the K-experiment were to study how product quality is influenced by the fertilization system and to develop quality-assessment methods. The K-experiment was designed with eight treatments and four crops each year but without replications. Variation in mean values was generally higher between years than between treatments, but the relation between treatments for each individual year was mostly the same for most of the response parameters. Still, it has been possible to compare the results of biodynamic and conventional treatments in the K-experiment with those from the above-described daughter projects. However, it should be kept in mind that the K-experiment was limited to comparing fertilization regimes. The UJ-experiments included a comparison between alternative and conventional farming systems with respect to crop rotation and pesticide use (Dlouhy,1981 and Pettersson, 1982.) ***Quantitative parameters.***

Tuber yield

The average tuber yield in the K-experiment was nearly the same in the organic and conventional treatments, whereas the yield was significantly lower in the unfertilized treatment (287 dt/ha). During years with higher precipitation, when conditions were more conducive for the mineralization of nitrogen in organic manure, organic treatments tended to outperform the conventional ones. In both UJ experiments, with shorter experiment periods, the yield was significantly lower (ca 20 percent) in the biodynamic treatments, partly owing to the higher yield losses caused by Phytophtora since pesticides were used in the conventionally fertilized treatments. This difference was partly compensated for by a better storability of the biodynamcially produced potatoes.

Dry matter content

There was a clear tendency for dry matter content to be higher in the biodynamic treatment than in the conventional one in both the K-experiment and in the two UJ-experiments. Dry matter content was significantly higher in both B1 and B2 compared with the conventionally fertilized systems A1 and A2.

Protein levels and protein quality

The level of crude protein was determined on the basis of Kjeldal-N, whereupon the pure protein was precipitated with CuSO4 and NaOH and expressed as per cent of the crude protein. In terms of crude levels of protein in percent of dry matter a clear gradient was found from low levels in the organically grown samples to high levels in the conventionally grown ones. The crude protein content was also significantly higher in the inorganic treatments in both UJ-experiments (figure 2), but the content of relatively pure protein was significantly higher in the biodynamic treatments than in the inorganic ones. The content of the free amino acids was measured by titration with feromol according to Sörensen. In the K-experiment the contents of these low molecular-weight, non-protein nitrogen compounds were lower in the organic treatment K1 than in K8 in all 19 studied years and lower in K1 than in K7 in 13 of the 19 years during which this parameter was studied during the experiment period. The higher protein quality of the organically grown crops was confirmed by comparing the relative content of essential amino acids and the biological value of protein, expressed as an EAA-index value (Dlouhý, 1981; Pettersson, 1982).

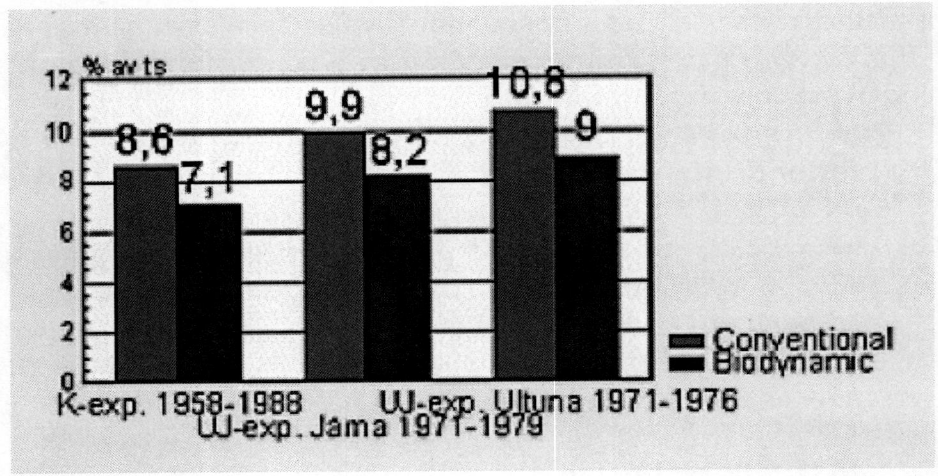

Figure 2. Potatoes. Crude protein in percent of dry matter. ***Physiological parameters.***

Darkening of tissue and extracts

The enzymatic darkening of raw, exposed potato tissue was measured with a reflectance attachment on a photospectrometer (described by Pettersson 1982). The speed of darkening of potato-tissue extracts was measured photometrically daily for four days at 530 nm. Both methods are described by Pettersson and E. v. Wistinghausen (1979) and Pettersson (1982). In the K and UJ-experiments both methods were used for evaluating the browning of the potatoes. The darkening of extract was greater in K1 than in K8 in 18 of the studied 24 years and greater in K1 than in K7 in 17 of the years. The UJ-experiments revealed that discoloration was more pronounced and developed faster in the chemically-fertilized variants than in the organically fertilized ones, and the difference was significant ($P<0.01$ in the UJ-experiments).

Extract decomposition

In this particular test of the resistance against the enzymatic and bacterial decomposition in water extract (Rd/Ro = the maximum decrease in the electrical resistance, in percent of the starting value in extract dissolution 1:10 during 4-5 days, according to Pettersson, 1982), decomposition values were generally lower for the organically grown variants than for the others (24 of 24 studied years in K1 compared with K8 but only in 15 of 24 years in K1 compared with the K7). Differences similar to those obtained between K1 and K8 were found in the comparison of conventional and biodynamic treatments in the UJ-experiments (P<0.01 in the UJ-experiments).

Storage losses

Storage tests were conducted harvesting in? April in 20-kg bags. The storage losses was measured in percent of the initial weight and included losses through respiration and damage caused by storage fungi, etc. In the K-experiment storage losses tended to be lower in the organic treatments than in the inorganic ones (9 of studied 11 years). This difference was even more pronounced in the UJ experiment at Järna and Uppsala (Figure 3) (P<0.05 and P<0.1).B

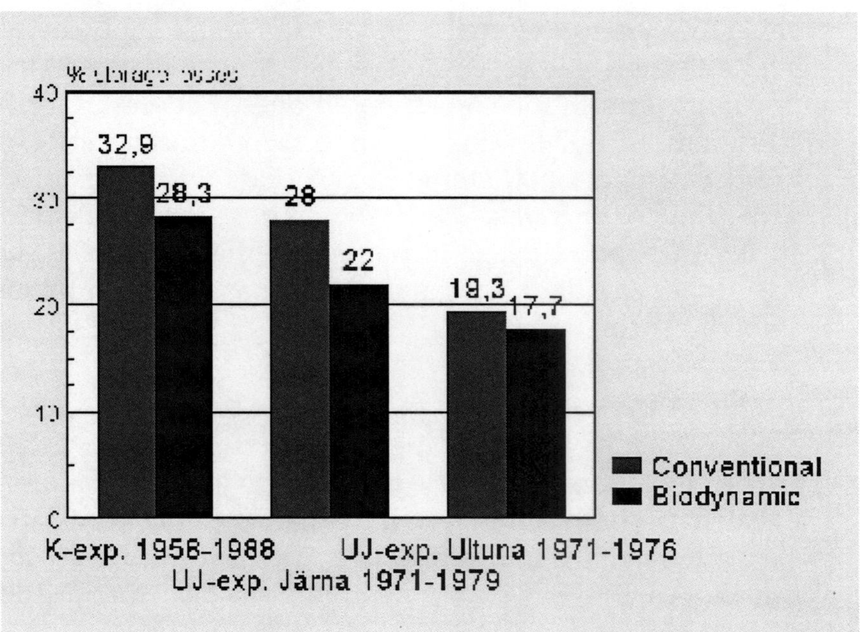

Figure 3. Potatoes. Storage losses after 6-7 months in percent of original stored weightField survey of Phytophthora infestans

Frequencies of infection were significantly lower in treatments K1 and K2 (composted manure) and in K5 as compared with K8 (high rate of NPK), and K3 and K4 (raw manure) over the 14-year period during which this parameter was studied. In the UJ-experiments the conventional treatments were treated with fungicide, and the yield difference was highest during the years with high infection levels (4 years).

Pathogen infection with Phytophora infestans (studied only during 1966-1969)

Although the spread of values was great with this testing method, a similar trend in the results was, nonetheless, seen, with lower values (i.e. less in vitro infection) in the organically grown samples than in the conventionally grown ones.

Morphological methods. Morphology of stems

The number of horizontal stems did differ appreciably between treatments, although values tended to be somewhat lower in the organic variants than in the conventional ones (Pettersson, Brinton & v, Wistinghausen, 1979). This negative correlation indicates that a low number of horizontal stems corresponds with high product quality according to the index. The method is described by Pettersson (1970). In the UJ-experiment at Järna (where this method was also used) the number of side (horizontal) stems was significantly lower in the biodynamic treatments.

Crystallization investigation

As with the foregoing results, these tests (Engqvist, 1970; Pettersson, 1982) revealed a similar trend, with organizational traits in the tissues being better in the organically grown samples than in the conventional ones. In the K-experiment, this was true for all studied years between 1966 and 1989.

Influence of previous crop on the quality of potatoes

The effects of ley and barley on quality parameters of succeeding crops differed in some respects. For example, potatoes following a ley tended to show a higher degree of extract dissolution and to have a higher nitrate content compared with potatoes following barley in the UJ-experiment. This type of farming-system effect was not possible to study in the K-experiment which was strictly a fertilization experiment.

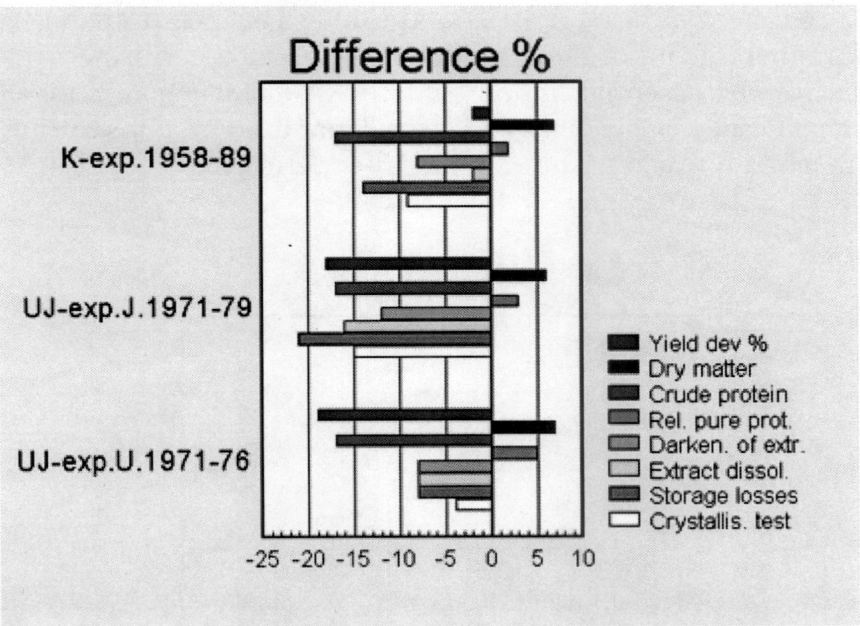

Figure 4. Potatoes. K-experiment 1958-1989, UJ-experiment in Järna 1971-1979, UJ-experiment in Uppsala 1971-1976. Difference, in percent, between the biodynamic treatments and the conventional ones for parameters.. Wheat in the K- and UJ-experiments

Average yield

Average yield levels in K1 and K8 were nearly the same. For K2, in which no biodynamic field preparation was carried out, the yield level was significantly lower, and differences were highest during years with a low yield level. In the UJ-experiments, with shorter experimental periods, the yield levels were significantly lower in the biodynamic treatments.

Quality parameters

In wheat as well, the crude protein content was higher in the inorganic treatments in the K-experiment and in both UJ-experiments (Figure 5). However, the content of relatively pure protein was higher in the biodynamic treatments in the K-experiment and higher in the UJ-experiment at Järna ($P<0.01$ and $P<0.1$) That the protein quality was higher in the organic treatments was also confirmed by the index for the essential amino acids (EAA-index, Figure 5 a), which was significantly higher in the biodynamic manured systems during the years when it was measured in the UJ-experiment.

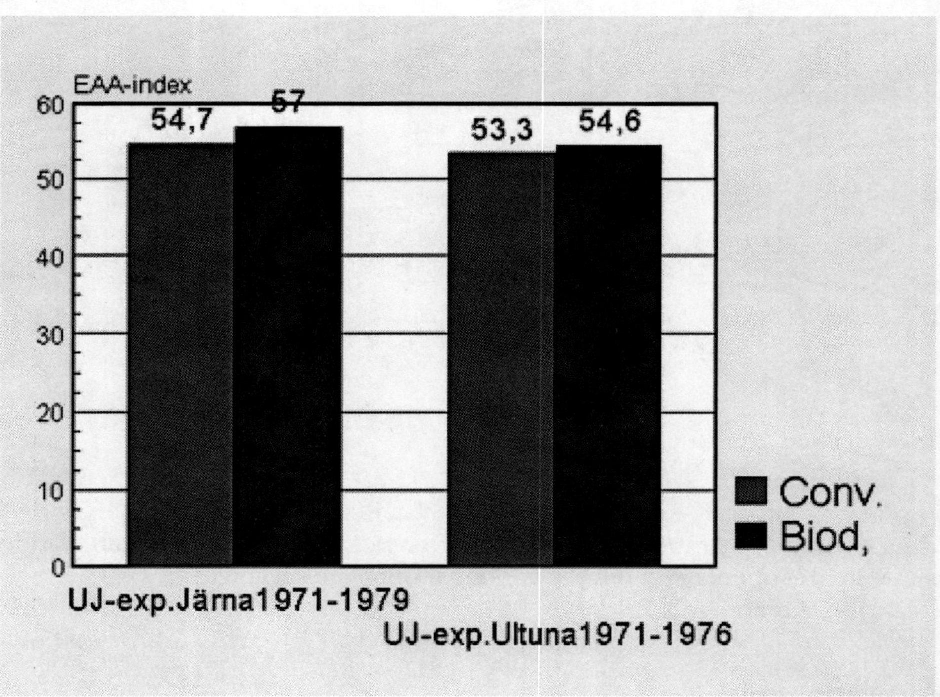

a

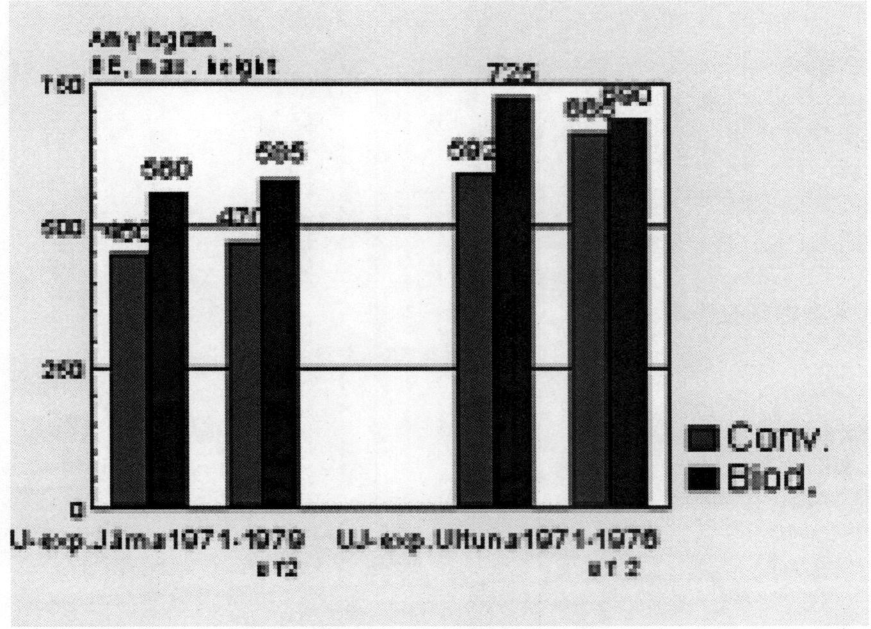

b

Figure 5 a. Index of essential amino acids in the UJ-experiments. Figure 5 b. Amylase activity in wheat in the UJ-experiments.

The resistance against extract dissolution

was also higher in the biodynamic treatments in these studies. In addition, starch quality seemed to be higher in the biodynamic treatments, measured in terms of falling number (in the K-experiment and the UJ-experiment in Järna) and as indicated by amylograms (figure 5 b). The differences, in percent, between the biodynamic treatments and the conventional ones for quantitative and qualitative parameters for wheat are illustrated in Figure 6.C

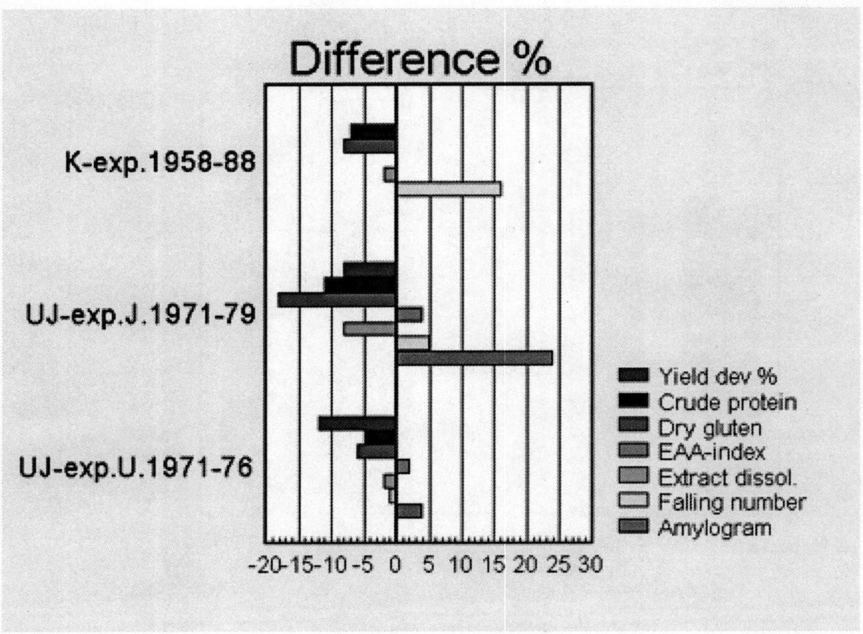

Figure 6. Wheat. K-experiment, 1958-1989, UJ-experiment in Järna, 1971-1979, UJ-experiment in Uppsala 1971-1976. Difference, in percent, between the bio-dynamic treatments (K1) and the conventional ones (K8) for quantitative and qualitative parameters for wheat Final discussion concerning the significance of the studied quality parameters.

Correlation between crude protein content and other parameters

In the UJ experiment at Järna significant correlations were observed between crude protein content and most of the studied parameters in both potatoes and wheat (Pettersson, 1982). For potatoes this parameter was correlated with dry matter content (negative, P<0.001), relatively pure protein (negative, P<0.01), EAA-index (negative, P<0.001), ascorbic acid content (negative, P<0.001), cooking quality (negative, P<0.01), taste quality (negative, P<0.001), free amino acids (positive, P<0.001), extract dissolution (positive, P<0.001) and darkening of extract (positive, P<0.001). For wheat, the correlations were with EAA-index (negative, P<0.01), extract dissolution (positive P<0.001) falling number (positive P<0.05), gluten content and baking tests (positive, $P<0.01$). ***It can thus be concluded that cereals with a high crude protein content, which is generally desired by the baking industry, tend to rank low in terms of biological and physiological quality parameters. Similarly potatoes with a high crude protein content tend to rank low in terms of biological and physiological quality parameters and taste.***

Light and shade

The studied parameters were divided into three groups: chemical/ biological, physiological and a third called morphological. Parameters falling into the last-mentioned group has been studied through a method based on observations of the morphology of stems in potatoes and a picture formation method (Crystallization with CuCl2). It was assumed that the lower quality of the crops, assessed in terms of these parameters studied here, in the conventionally fertilized systems is similar to what would have been found had the plants been grown under more shaded conditions. Likewise, the higher biological value of the protein and the higher values found for quality parameters are similar to what would have been expected had the plants been grown under full light. In the interpretation, the organic treatments are described as having the same effect as an increase in solar radiation, and likewise, the effects of NPK-treatments are analogous to those induced by an increase in shade (Pettersson, 1982).

"Organization level"

The better protein composition, with a lower content of free amino acids, and better storage properties were attributed to the higher organization level resulting from the organic and biodynamic fertilizing systems. The higher "organization level", reflected in the higher biological value of the protein and lower content of free amino acids not used in protein synthesis, was correlated with better resistance against pests, less darkening, less extract dissolution and lower losses of potato during storage. Future research should be aimed at improving our understanding of the significance of these qualitative values in terms of their effects on the human health.

Further studies

Upon completion of the K-experiment, a new experiment was begun in both Sweden (1992) and Finland (1995) in which different manure systems were compared in connection with various types of crop rotations on organic farms. The aim is to gain a better understanding of the differences between composted, non-composted and liquid manure in terms of their effects on the parameters discussed here and to develop effective methods for regularly testing product quality. In addition, the biodynamic preparation treatments are being studied so that their effects can be predicted more reliably. Preliminary results from studies in potatoes in Finland indicate that the use of liquid manure, like mineral fertilizer, tends to decrease the biological value. The ultimate goal is to tailor the fertilizing regime to the soil, climate and crops in a way that offers both a good yield an a healthy product with a high nutrient value.

REFERENCES

1. Dlouhy, J. 1981. Alternativa odlingsformer - växtprodukters kvalitet vid konventionell och biodynamisk odling (Alternative forms of agriculture - quality of products from conventional and biodynamic growing. With English summary). Swedish University of Agricultural Sciences. Dep. of Plant Husbandry. Report 91. Uppsala.
2. Engqvist, M. 1970. Gestaltkräfte des Lebendigen. Vittorio Klostermann, Frankfurt am Main.
3. Granstedt, A. 1992. The potential for Swedish farms to eliminate the use of artificial fertilizers. American Journal of Alternative Agriculture. Vol 6. Numb. 3, 122-131.
4. Granstedt, A. 1995. The mobilization and immobilization of soil nitrogen after green manure crops. I: Proceedings of the third international conference, Soil management in sustainable agriculture, 31 August - 4 September 1993. Wye College. University of London, 265-275
5. Pettersson B.D. 1970. Verkan av växtplats, gödsling och tillväxtreglerande substanser på matpotatisens egenskaper. (Influence of growing locality, manure and growth regulating substances on the quality properties of potatoes Nordisk Forskningsring. Meddelande nr 23, Järna.
6. Pettersson, Brinton & v, Wistinghausen, E.v. 1979. Effects of organic and inorganic fertilizers on soils and crops. Results of a long term field experiment in Sweden. Nordisk Forskningsring, Meddelande Nr. 30. Järna.
7. Pettersson, B. D. 1982. Konventionell och biodynamisk odling. Jämf`rande försök mellan tvD odlingssystem. (Investigations within Conventional and Biodynamic Farming Systems. Summary in English). Nordisk forskningsring, Järna.

8. Pettersson, B. D., Reents, H. J. & Wistinghausen E.v. 1992. Düngung und Bodeneigenschaften - Ergebnisse eines 32-jahrigen Feldversuches in Järna, Schweden.Nordisk Forskningsring, Meddelende nr 34. Järna.

Chapter 14

MICROBIAL MINERALIZATION OF ORGANIC NITROGEN INTO NITRATE TO ALLOW THE USE OF ORGANIC FERTILIZER INHYDROPONICS

Makoto Shinohara [a] , Chihiro Aoyama [b] , Kazuki Fujiwara [c] , Atsunori Watanabe [b] , Hiromi Ohmori [a] , Yoichi Uehara [a] & Masao Takano [b]

[a] National Institute of Vegetable and Tea Science, National Agricultural Research Organization (NARO) , Kusawa 360, Ano, Tsu , Mie 514-2392

[b] Nagoya University, Frou, Chigusa , Nagoya, Aichi, 464-8601 , Japan

[c] Royal Agricultural College, Cirencester , Gloucestershire GL7 6JS , UK

ABSTRACT

Hydroponics is an excellent technique for the cultivation of vegetable crops and other plants, but organic fertilizers cannot be used in conventional hydroponic systems, which generally use only inorganic fertilizers, because organic compounds in the hydroponic solutions generally have phytotoxic effects that lead to poor plant growth. Few microorganisms are present in hydroponic solutions to mineralize the organic compounds into inorganic nutrients. In this article a novel and practical hydroponic culture method that uses microorganisms to degrade organic fertilizer in the hydroponic solution has been developed. Soil microorganisms were cultured by regulating the amounts of organic fertilizer and inoculum, with moderate aeration. The microorganisms mineralized organic nitrogen via ammonification and nitrification into nitrate at efficiency of 97.6%. The culture solution containing the microorganisms was usable as a hydroponic solution, and organic fertilizer could be directly added to it during vegetable cultivation. Vegetables grew well in the organic hydroponic system. Organic hydroponics based on this method is therefore a practical tool for the utilization of organic sources of fertilizer

INTRODUCTION

Hydroponic culture promotes the growth of crops by allowing close control of fertilization. A closed recirculating hydroponic system conserves fertilizer and water; in contrast, soil culture represents an open system with a relatively low efficiency of water and fertilizer use (Hagin and Lowengart 1996). Conventional hydroponic systems cannot use organic fertilizer, which inhibits plant growth: organic compounds contained in hydroponic solutions have been regarded as phytotoxic (Garland and Mackoqiak 1990; Garland et al. 1993; Mackowiak et al. 1996; Garland et al. 1997; Atkin and Nichols 2004; Ehret et al. 2005; Lee et al. 2006). However, from the viewpoint of resource recycling, it is important to develop methods capable of using organic fertilizer sources in hydroponics.

The ability to use organic fertilizer in hydroponics ("organic hydroponics") has been studied as a method to grow crops in space habitats (Garland and Mackoqiak 1990; Garland et al. 1993; Mackowiak et al. 1996; Stutte 1996; Garland et al. 1997). However, the direct use of organic fertilizer proved to be deleterious to plant growth (Garland et al. 1997). Therefore, organic fertilizer has been microbially pre-processed before incorporation into hydroponic solutions (Strayer et al. 1997; Atkin and Nichols 2004). The pre-processing has usually been conducted in two or more reaction tanks to provide ammonification and nitrification of organic nitrogen compounds. However, the efficiency of generating nitrate from organic nitrogen in the organic fertilizer was less than 30% (Strayer et al. 1997). Nitrate is the desired form of nitrogen source for crop production, because many vegetable crops prefer this form over ammonium nitrogen (Ikeda and Osawa 1981; Strayer et al. 1997; Miyata and Ikeda 2005). Ammonium can be used as part of a mixed nitrogen source in hydroponic solutions, but crop failures are very real possibilities if ammonium is the primary nitrogen form (Trebst et al. 1960; Purich and Barker 1967; Givan 1979; Ikeda and Osawa 1981; Stutte 1996; Miyata and Ikeda 2005). For practical and successful organic hydroponics, it is necessary to efficiently generate nitrate from organic fertilizer, thereby allowing direct addition of organic fertilizer to the hydroponic solution during cultivation.

In soil culture, organic fertilizer is usually added directly during cultivation. The organic fertilizer incorporated into the soil is degraded by soil microorganisms, which generate nitrate via ammonification and nitrification, and this nitrate is absorbed rapidly by plants. If the microbial community that degrades organic fertilizer can be cultured in hydroponic solution, it should be possible to add organic fertilizer directly to the solution. However, past studies of organic hydroponics have not tried this.

In this paper, we describe fine-tuning of a previously developed method that can generate nitrate efficiently by culturing microorganisms capable of degrading organic fertilizer in the hydroponic solution (Shinohara 2006, 2007). Vegetables can then be grown in the resulting hydroponic solution. Furthermore, we successfully grew lettuce and tomato plants in the hydroponic solution by means of direct addition of organic fertilizer to the solution during cultivation. Fish-based soluble fertilizer, a by-product of the production of dried bonito fish flakes, was the main organic fertilizer.

MATERIALS AND METHODS

Generation of Nitrate from Organic Fertilizer by Microbial Degradation in Water

To find an appropriate source of microorganisms that could mineralize organic fertilizer, we added 5 g/L of field soil collected from a field of garden pea, Pisum sativum, at Japan's National

Institute of Vegetable and Tea Science. We also tested commercial "Naeichiban" nursery soil (Sumirin-Nousankougyo Co., Ltd, Kaifu, Aichi, Japan) and "Golden bark" compost (Shimizu-kou Mokuzai-Sangyou Kyoudou Kumiai, Shizuoka, Shizuoka, Japan) as sources of inoculum. The soils were added to 100-mL flasks (n¼3 per treatment) of distilled water containing 1 g/L of fish-based soluble fertilizer (Makurazaki Gyokyo-Kumiai, Makurazaki, Kagoshima, Japan). As a control, we used distilled water containing the fish-based soluble fertilizer without inoculum. The fertilizer contained 4.5 mol/kg nitrogen (N), 0.019 mol/kg phosphorus (P), and 0.23 mol/kg potassium (K). We also tested 100-mL samples of sea water collected from Kinuura Harbor in Aichi Prefecture as inoculum, without soils or bark compost. The flasks were shaken (120 strokes/min) for 20 days at 25°C, and the nitrate and ammonium concentrations were then determined.

Based on the results of the previous experiment, we chose bark compost as the inoculum in the next experiment. To determine the optimum dose of inoculum, we added bark compost at 0, 0.5, or 5 g/L to 100- mL flasks (n¼3 per treatment) of distilled water containing 0.25 g of fish-based soluble fertilizer or corn steep liquor (CSL; Sakata, Yokohama, Kanagawa, Japan), which contained 2.4 mol/kg N, 0.48 mol/kg P, and 0.69 mol/kg K; the net concentration of fish-based soluble fertilizer or CSL was 2.5 g/L. The flasks were shaken (120 strokes/min) for 21 or 13 days, respectively, at 25°C, and then the nitrate and ammonium concentrations were determined.

To determine the optimum dose of organic fertilizer, we added fish-based soluble fertilizer at 0.5, 2.5, or 5 g/L daily for seven days from the start of the experiment to 2 L of water containing 5 g/L bark compost as microbial inoculum. The experiment was performed for 15 days at an ambient temperature of 25°C in buckets (n¼3 per treatment); during this time, the water was aerated (19.6 kPa) with a Nisso α -4000 aeration pump (Marukan Co. Ltd, Kasukabe, Saitama, Japan).

To determine the efficiency of conversion of organic nitrogen into nitrate, we added 1 g/L of canola oil cake, corn oil cake, CSL, soybean curd refuse, low-grade spirits distilled from sake lees, rice bran, fermented poultry manure, or fish flour to 50-mL flasks (n¼3 per treatment) of distilled water containing 0.5 g of bark compost.

The flasks were shaken (120 strokes/min) for 17 days at 25°C, and the ammonium and nitrate concentrations were then determined.

In all these analyses, we used an RQ-Flex Plus Analyser (Merck, Frankfurt, Germany) to determine the concentrations of nitrate and ammonium ions. Microsoft Office Excel 2003 was used as the statistical software.

Vegetable Growth

Growth of Tomato Seedlings with the Microbial Culture Solution as a Hydroponic Solution

We conducted growth experiments with tomato seedlings using the microbial culture solution as the hydroponic solution. We raised four tomato (Solanum lycopersicum cv. "Ponderosa"; Kobayashi Seed Co., Ltd, Kakogawa, Hyogo, Japan) seedlings in a pot (ø 23 cmheight 23.2 cm) per replicate (n¼3 replicates per treatment). To prepare the microbial culture solution, we added fishbased soluble fertilizer at 300 g per 200 L of water containing 200 g of bark compost as a microbial inoculum, then aerated the water (19.6 kPa) for 50 days using two aeration pumps (Nisso α-4000). The prepared microbial culture solution contained 123 mg/L nitrate and no ammonium. Each hydroponic pot was filled with 6 L of a solution containing 11.9 mmol nitrate-nitrogen. As the control treatment, we used the conventional inorganic hydroponic solution, which contained 1 g of Otsuka House TM No. 1 and 0.67 g of Otsuka House TM No. 2 (Otsuka Chemical Co., Ltd, Osaka, Japan). This provided the following nutrients: 11.9 mmol N (1.8% ammonium-nitrogen and 98.2% nitrate-nitrogen), 1.07 mmol P, 5.5 mmol K, 0.91 mmol magnesium (Mg), 0.013 mmol manganese (Mn), 0.013 mmol boron (B), 0.03 mmol iron (Fe), and 2.63 mmol calcium (Ca). We also added 60 g of oyster shell lime (Urabe Industry Co., Ltd, Fukuyama, Hiroshima, Japan) suspended in each pot to provide additional nutrients and minor nutrients: 11.9 mmol Mg, 9.67 mmol Fe, 2.17 mmol B, 1.53 mmol Mn, 0.077 mmol zinc (Zn), 0.0113 mmol copper (Cu), 0.005 mmol molybdenum (Mo), and 62.7 mmol Ca. We added fish-based soluble fertilizer at 3 g/pot (i.e. 0.5 g/L) daily for three days from the start of the experiment to the hydroponic solution in each pot. During cultivation, the solution was aerated (19.6 kPa) with a Nisso α-4000 aeration pump. The experiments were conducted in a glasshouse at Tsu, Mie Prefecture, July 21–29, 2010. The hydroponics systems using microbial and nonmicrobial solutions were each established in three pots per treatment.

Growth of Lettuce with the Microbial Culture Solution as a Hydroponic Solution

We also conducted growth experiments with butterhead lettuce seedlings using the microbial culture solution as the hydroponic solution. We raised four butterhead lettuce "Saradana" (Lactuca sativa var. capitata; Sakata Seed Corporation, Yokohama, Kanagawa, Japan) seedlings in a pot (width 9 cm×length 9 cm×height 15 cm) per replicate (n¼3 replicates per treatment). To prepare the microbial culture solution, we added fish-based soluble fertilizer at 300 g per 200 L of water containing 200 g of bark compost as a microbial inoculum, then aerated the water (19.6 kPa) for 170 days using two aeration pumps (Nisso α-4000). The prepared microbial culture solution contained 210 mg/L nitrate and 8.3 mg/L ammonium. Each hydroponic pot was filled with 1 L of a solution containing 3.98 mmol nitrate-nitrogen. As a control treatment, hydroponic solution was prepared with 1 L of distilled water containing 1 g/L of fish-soluble without inoculation of the microbial culture solution.

The hydroponic solution contained 60 mg/L organic nitrogen. Each hydroponic pot was filled with 1 L of a solution containing 4.29 mmol organic nitrogen. We added 10 g of oyster shell lime (Urabe Industry Co., Ltd, Fukuyama, Hiroshima, Japan) suspended in each pot to provide additional nutrients and minor nutrients: 1.98 mmol Mg, 1.61 mmol Fe, 0.36 mmol B, 0.255 mmol Mn, 0.013 mmol Zn, 0.00188 mmol Cu, 0.0008 mmol Mo, and 10.5 mmol Ca. As another control treatment, conventional inorganic solution was used as a starter of hydroponic solution, which contained 1 g of Otsuka House TM No. 1 and 0.67 g of Otsuka House TM No. 2 (Otsuka Chemical Co., Ltd, Osaka, Japan). This provided the following nutrients: 3.98 mmol N (98.2% nitrate-nitrogen and 1.8% ammonium-nitrogen), 0.358 mmol P, 1.84 mmol K, 0.30 mmol Mg, 0.0043 mmol Mn, 0.0043 mmol B, 0.010 mmol Fe, and 0.88 mmol Ca. We added fish-based soluble fertilizer at 1 g/pot (i.e. 1 g/L) at the start of the experiment to the hydroponic solution in each pot. During cultivation, the solution was aerated (19.6 kPa) with a Nisso α-4000 aeration pump. The experiments were conducted in a glasshouse at Tsu, Mie Prefecture, November 12–29, 2010. The hydroponics systems using microbial and nonmicrobial solutions with or without conventional inorganic hydroponic solution were each established in three pots per treatment.

Growth of Tomato Seedlings with Oyster Shell Lime

We conducted growth experiments with the tomato (cv. "Ponderosa") seedlings, with or without oyster shell lime as a supplement to provide minor nutrients. We raised four seedlings in a pot (ø 23 cm×height 23.2 cm) per replicate (n¼3 replicates per treatment). To prepare the hydroponic solution, we added fish-based soluble fertilizer at 300 g per 200 L of water containing 200 g bark compost as a microbial inoculum, then aerated the water (19.6 kPa) for 33 days using two aeration pumps (Nisso α-4000). The prepared hydroponic solution contained 164 mg/L nitrate ions and 9 mg/L ammonium ions. Each hydroponic pot was filled with 5 L of water and 1 L of a solution containing 3.8 mmol nitrogen as nitrate. We added 0 or 60 g of oyster shell lime (Urabe Industry Co., Ltd) suspended in the solution to provide additional nutrients and minor nutrients: 11.9 mmol Mg, 9.67 mmol Fe, 2.17 mmol B, 1.53 mmol Mn, 0.0771 mmol Zn, 0.0113 mmol Cu, 0.005 mmol Mo, and 62.7 mmol Ca. During cultivation, the solution was aerated (19.6 kPa) with a Nisso -4000 aeration pump. The experiments were conducted in a glasshouse at Tsu from August 12 to September 2, 2010. We added 0.12 g ethylenediaminetetraacetic acid iron salt (FeEDTA) (Dojindo Laboratories, Kumamoto, Japan) in a pot as an Fe supplement eight days after the start of the experiment.

Lettuce

We conducted growth experiments with butterhead lettuce "Saradana" (Lactuca sativa var. capitata; Sakata Seed Corporation, Yokohama, Kanagawa, Japan) in a glasshouse at Tsu from October 15 to November 9, 2009. During these experiments, light and temperature were not controlled. We raised 10 butterhead lettuce seedlings in a polystyrene foam board floating in one hydroponic container (length 50width 20×height 25 cm) per replicate (n¼3 replicates per treatment). To prepare the hydroponic solution, we added fish-based soluble fertilizer at 300 g per 200 L of water containing 200 g of bark compost as a microbial inoculum; the water was aerated (19.6 kPa) for a total of 48 days before cultivation using two aeration pumps (Nisso α-4000). The prepared hydroponic solution contained 235 mg/L of nitrate but no ammonium. Each hydroponic container was filled with 15 L of a solution containing 56.9 mmol nitrogen as nitrate. We also added 150 g of oyster shell lime (Urabe Industry Co., Ltd) suspended in the solution to provide additional nutrients and minor nutrients: 29.8 mmol Mg, 24.2 mmol Fe, 5.41 mmol B, 3.82 mmol Mn, 0.193 mmol Zn, 0.0283 mmol Cu, 0.0125 mmol Mo, and 157 mmol Ca.

During cultivation, the solution was aerated (19.6 kPa) with a Nisso α-4000 aeration pump. As the control treatment, we used conventional hydroponics with inorganic fertilizer containing 4.78 g of Otsuka House TM No. 1 and 3.18 g of Otsuka House TM No. 2, which provided the following nutrients: 56.9 mmol N (1.8% ammonium-nitrogen and 98.2% nitrate-nitrogen), 5.1 mmol P, 26.3 mmol K, 4.35 mmol Mg, 0.06 mmol Mn, 0.063 mmol B, 0.14 mmol Fe, and 12.6 mmol Ca. We adjusted the amount of fertilizer added to the system to provide the same amount of nitrogen as was provided by the organic hydroponic system. To determine the concentration of nutrients in the hydroponic solutions the next day at the start of the experiment, the solutions were analyzed by inductively coupled plasma atomic emission spectroscopy (ICP-AES) SPS-7700 (Seiko Instruments Inc. Chiba, Chiba, Japan). The organic and conventional hydroponics were each performed in three containers.

Tomato

We used a nutrient film technique (Wheeler et al. 1990) to grow cv. "Saturn" tomato plants (Takii & Co., Ltd, Kyoto, Japan; n¼24 per treatment) in a closed recirculating hydroponic system (M-shiki Suikou Co., Ltd, Yatomi, Aichi, Japan; length 465×width 60×height 18 cm) based on a deep-flow technique from June 7 to October 13,

2006. This experiment was conducted in a glasshouse at Taketoyo, Aichi Prefecture.

To prepare the hydroponic solution, we added fish-based soluble fertilizer at 100 g or CSL at 200 g per 200 L of water containing 200 g bark compost as a microbial inoculum; the hydroponic solution was aerated with an MD-15 R-N water pump (Iwaki Co. Ltd, Tokyo, Japan) by recirculation of the nutrient solution in the hydroponic system for 12 days. The nitrate concentrations in the starter hydroponic solutions were 189 mg/L (fish-based fertilizer) and 420 mg/L (CSL); no ammonium-nitrogen was detected in either hydroponic solution. At the start of cultivation we also added 560 g of oyster shell lime to each hydroponic solution to provide minor nutrients and micronutrients: 111 mmol Mg, 90.2 mmol Fe, 20.2 mmol B, 14.3 mmol Mn, 0.720 mmol Zn, 0.106 mmol Cu, 0.047 mmol Mo, and 585 mmol Ca . For 24 days from the start of cultivation, we added fish-based soluble fertilizer at 0.5 g/plant daily or CSL at 1 g/plant daily to provide the equivalent of 30 mg N/plant daily; we subsequently added fish-based soluble fertilizer at 1 g/ plant daily or CSL at 2 g/ plant daily to provide the equivalent of 60 mg N/plant daily. The organic fertilizers were added directly to each hydroponic solution. We also added wood ash (JOY AGRIS Co., Ltd, Tokyo, Japan) containing 1.64 mol/kg K to the hydroponic solution as a potassium supplement during cultivation, at 0.34 g (containing 0.56 mmol K) per 1 g of CSL or 0.932 g per 1 g of fish-based soluble fertilizer. In the control treatment, we conducted conventional hydroponics with inorganic fertilizer solution that comprised a mixture of Otsuka House TM Nos 1, 2, and 5, at proportions of 30:20:1, respectively. This provided the following nutrient concentrations: 18.8 mmol/L N (1.8% ammoniumnitrogen and 98.2% nitrate-nitrogen), 1.69 mmol/L P, 8.69 mmol/L K, 1.49 mmol/L Mg, 0.04 mmol/L Mn, 0.04 mmol/L B, 0.1 mmol/L Fe, and 4.1 mmol/L Ca.

The amount of fertilizer added to the system was adjusted to provide the same amount of nitrogen as in the organic hydroponic system. The concentration of nitrate ion in every hydroponic solution was daily analyzed just before addition of fertilizers in the hydroponic solution. The Brix values (a measure of the sugar content) of the fruits were measured in supernatant filtered from homogenized fruits, using an APAL-1 Refractometer (ATAGO Co.,

Ltd, Tokyo, Japan). We used an RQ-Flex Plus Analyser to quantify the ascorbic acid contents.

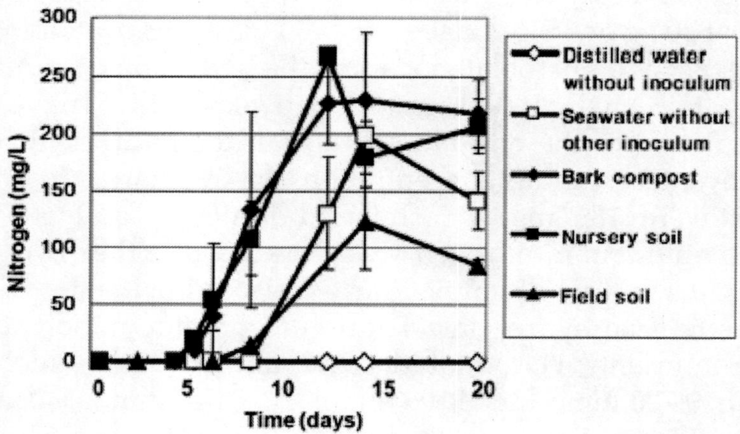

Figure 1. Mineralization of organic nitrogen into nitrate by the addition of microbial inocula from various sources. The suitability of various sources of inoculum of microorganisms capable of mineralizing organic fertilizer into nitrate was examined: 5 g/L of field soil, nursery soil, or bark compost was added as microbial inoculum to 100mL of distilled water containing fish-based soluble fertilizer at 1 g/L. Sea water was also substituted for distilled water, but without other inoculum. Distilled water without inoculum was used as the control. Values represent the means and standard deviations (n¼3).

RESULTS

Generation of Nitrate from Organic Fertilizer by Microbial Degradation in Water

We chose field soil, nursery soil, bark compost, and sea water as inoculum sources for the microorganisms needed to mineralize organic nitrogen into nitrate in water in the presence of fish-based soluble fertilizer as a nitrogen source. The addition of nursery soil or bark compost as inoculum sources generated nitrate (Fig. 1). Sea water without soil or bark compost but with added fish-based soluble fertilizer also generated nitrate. The efficiency of generation did not differ significantly ($P > 0.05$) among the inocula; however, all three inocula were significantly ($P < 0.05$) more efficient than field soil.

However, distilled water without inoculum but with added fish-based soluble fertilizer generated ammonium but no nitrate. These results indicate that the saprophytic bacteria in the fertilizer could generate only ammonium from the organic nitrogen in the fish-based soluble fertilizer, but microorganisms added to the water from the field soil, nursery soil, bark compost, or sea water were able to mineralize the organic nitrogen into nitrate (Fig. 1). We performed all subsequent experiments using bark compost, because it provided good results and is easily obtainable.

We examined the inoculum requirements with bark compost in 100-mL flasks of distilled water containing 0.25 g CSL (Fig. 2). The nitrate content increased as the ammonium content decreased in the solution with bark compost at 5 g/L, but no nitrate was detected in flasks when 0.5 g/L or less of bark compost was added. These results suggest that about 5 g/L of bark compost inoculum was needed to mineralize organic compounds into nitrate. However, nitrate was detected in flasks containing 0.25 g of fish-based soluble fertilizer per 100mL when 0.5 g/L or more of bark compost was added (data not shown). These results suggest that the organic nitrogen in fish-based soluble fertilizer can be more efficiently transformed into nitrate than that in CSL.

We then examined the optimum dose of organic fertilizer (Fig. 3). We added fish-based soluble fertilizer daily for seven days from the start of incubation to water containing 5 g/L of bark compost inoculum. Nitrate was generated only in containers that received 0.5 g/L of fertilizer daily for seven days; higher application rates produced only ammonium. The efficiency of nitrate generation from the organic nitrogen was 97.6%. Nitrification was inhibited in containers that received fertilizer at 2.5 or 5 g/L daily for seven days. This suggests that nitrification can be inhibited by excessive amounts of organic fertilizer.

We then examined the efficiency of conversion of organic nitrogen into nitrate in various organic fertilizers. Nitrate was generated in each flask that contained CSL (at an efficiency of 99.7%), corn oil cake (91.3%), fish flour (77.7%), canola oil cake (74.5%), soybean curd refuse (73.5%), and fermented poultry manure (30.5%). However, no nitrate was generated in the flasks that contained low-grade spirits distilled from sake lees or rice bran. The carbon-to-nitrogen ratios (C/N) of these organic fertilizers were 0.90 for CSL, 3.45 for fish-based soluble fertilizer, 4.66 for fish flour, 6.93 for canola oil cake, 7.0 for fermented poultry manure, 8.68 for corn oil cake, 10.43 for soybean curd refuse, 11.78 for the sake lees, and 18.1 for rice bran.

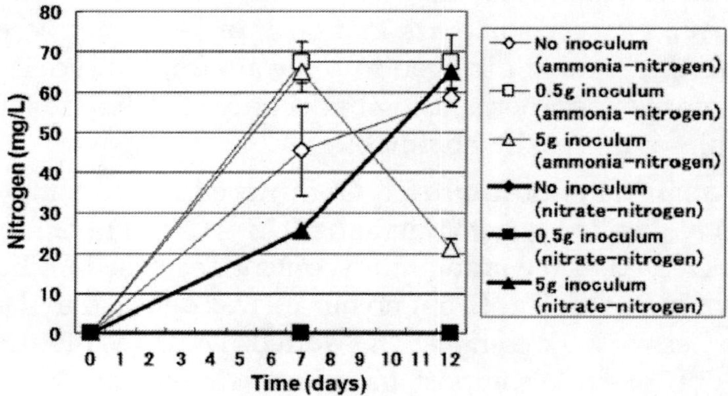

Figure 2. Inoculum requirements with bark compost. The optimum concentration of bark compost inoculum for mineralization of organic nitrogen from corn steep liquor (CSL) into nitrate was determined. Bark compost was added at 0, 0.5, or 5 g/L as a microbial inoculum to 100mL of distilled water containing 2.5 g/L of CSL. Values represent the means and standard deviations (n¼3).

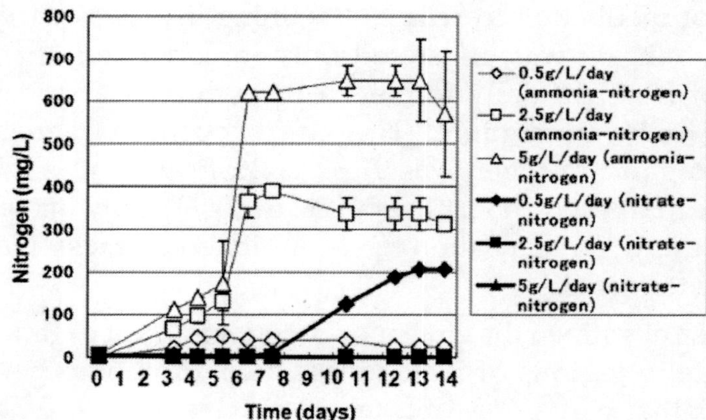

Figure 3. Optimum dose of organic fertilizer. The optimum concentration of organic fertilizer required to promote nitrification in the presence of 5 g/L of bark compost inoculum was determined. Fish-based soluble fertilizer at 0.5, 2.5, or 5 g/L was added daily for seven days from the start of the experiment. Values represent the means and standard deviations (n¼3).

Vegetable Growth

Growth of Tomato Seedlings with the Microbial Culture Solution as a Hydroponic Solution

The tomato seedlings grew well when we used the microbial culture solution as the hydroponic solution (Fig. 4a). In contrast, all the tomato seedlings grown without the microbial culture solution died (Fig. 4b). Nitrate was detected in the hydroponic solution with the microbial culture solution (206, 220, and 232 mg/L respectively, in the three media), but no ammonium was detected eight days after the start of the experiment. In contrast, only ammonium was detected (at 69, 70, and 80 mg/L respectively, in the three media) from the solutions that lacked any microbial culture solution.

Figure 4. Tomato growth experiment using microbial culture solution as a hydroponic solution. The growth of tomato plants grown with (a) the microbial culture solution or (b) conventional inorganic solution without microbial inoculation as a starter of the hydroponic solution was compared. Each pot was provided with 60 g of oyster shell lime. Fish-based soluble fertilizer was added at 3 g/pot daily for three days. These pictures were taken eight days after the start of cultivation.

Growth of Lettuce With the Microbial Culture Solution as a Hydroponic Solution

The lettuce seedlings grew well when we used the microbial culture solution (Fig. 5a). Similarly, the seedlings grew well in the system with conventional inorganic solution (Fig. 5c). In contrast, all the lettuce seedlings seriously wilted in the solution containing only the fish based soluble fertilizer five days after the start of the experiment (Fig. 5b).

Nitrate was detected in the hydroponic solution with the microbial culture solution (210, 230, and 235 mg/L respectively, in the three media) but no ammonium was detected seven days after the start of the experiment.

Nitrate was detected in the solution with conventional inorganic solution (85, 115, and 180 mg/L respectively, in the three media) and ammonium was generated (at 28, 28, and 28 mg/L respectively, in the three media). Only ammonium was detected (at 38, 31, and 34 mg/L respectively, in the three media) from the solutions added with only fish-based soluble fertilizer.

Growth of Tomato Seedlings with Oyster Shell Lime Provided

The tomato seedlings grown in the solution containing oyster shell lime grew well (Fig. 6a). In contrast, leaf tips of the seedlings grown without oyster shell lime became yellow and the growth decreased (Fig. 6b and c). The addition of 0.12 g FeEDTA to each pot ameliorated but did not eliminate the symptoms (Fig. 6d).

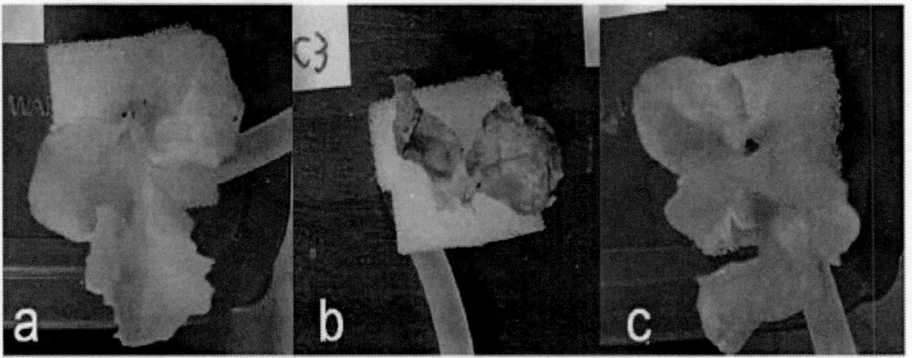

Figure 5. Lettuce growth experiment using microbial culture solution as a hydroponic solution. The growth of lettuce plants grown with (a) the microbial culture solution, (b) distilled water containing 1 g/L of fish-based soluble fertilizer, or (c) conventional inorganic solution as a starter of hydroponic solution was compared. Each pot was provided with 10 g of oyster shell lime. Fish-based soluble fertilizer was added at 1 g/pot at the start of the experiment. These pictures were taken five days after the start of cultivation.

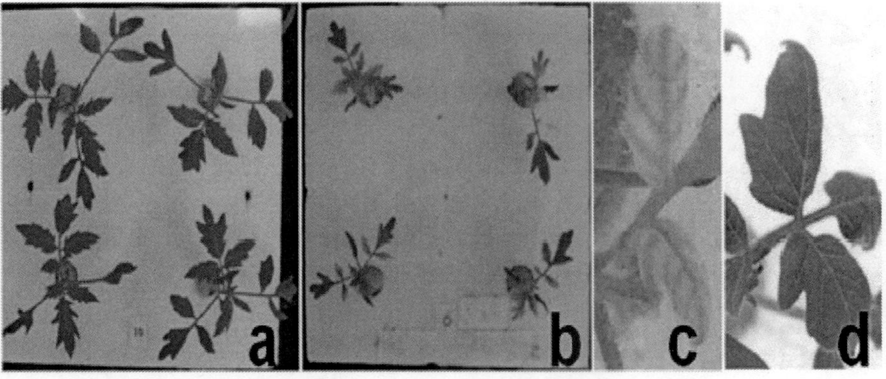

Figure 6. Tomato growth experiment using oyster shell lime to provide minor nutrients. The tomato seedlings were grown in the organic hydroponic system (a) with or (b) without 60 g of oyster shell lime. (c) In cultivation without oyster shell lime, the leaf tips of the seedlings turned yellow. (d) The addition of 0.12 g FeEDTA to the pot ameliorated but did not eliminate the symptoms.

Lettuce

We examined the growth of butter head lettuce in a hydroponic nutrient solution in which the organic nitrogen contained in fish-based soluble fertilizer was optimally mineralized into nitrate (Fig. 7). The organic system produced significantly greater ($P < 0.05$) fresh head weight and root dry weight than in the conventional system. (We measured head fresh weight instead of dry weight, because fresh weight determines the commercial value of the lettuce.) The leaf nitrate ion content was 35.5% lower in the organic system, and the difference was significant ($P < 0.05$). The hydroponic solutions, the next day of the start, of organic system and conventional system contained 28.82 and 35.44 mg/L K, 66.7 and 7.24 mg/L Na, 66.32 and 37.22 mg/L Ca, 0.4988 and 0.285 mg/L Fe, 10.1 and 12.6 mg/L Mg, 0.086 and 0.4048 mg/L Mn, and 0.0598 and 0.1674 mg/L B, respectively. Neither nitrate nor ammonium was detected in the hydroponic solution of either system at the end of cultivation. The ascorbic acid content of the leaves did not differ significantly ($P > 0.05$) between the two systems.

Tomato

We also performed a tomato growth study (Fig. 8). The total fruit yields (24 plants per treatment) were 35,338 g with the conventional fertilizer, 32,974 g with the CSL, and 33,783 g with the fish-based soluble fertilizer, and the fruit weight per plant did not differ significantly ($P > 0.05$). The average Brix value of the fruits (n¼32 fruits per treatment) in the conventional treatment (4.8) did not differ significantly from the value in the CSL treatment, but both were significantly lower ($P < 0.05$) than the value in the fish-based soluble fertilizer treatment (5.2). The average ascorbic acid content of the fruits (n¼32 fruits per treatment) did not differ significantly among the treatments.

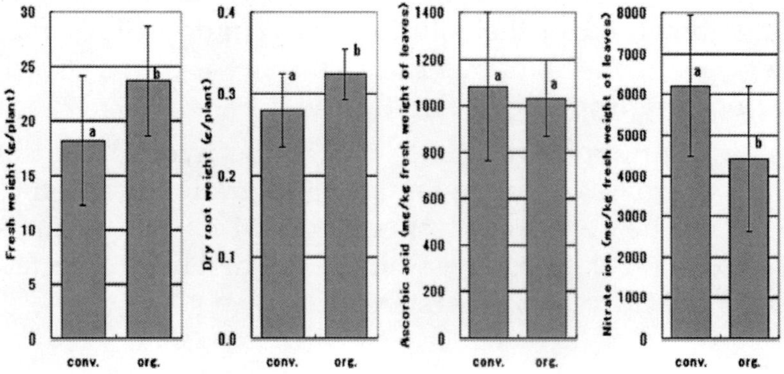

Figure 7. Butterhead lettuce growth in organic and conventional hydroponic systems. The growth and quality of butterhead lettuce plants grown using the organic hydroponic approach with fish-based soluble fertilizer (org.) and in conventional hydroponics with inorganic fertilizer (conv.) were measured. Values represent means and standard deviations (n¼30). Bars labeled with different letters differ significantly (Fisher's protected least-significant-difference test, P < 0.05).

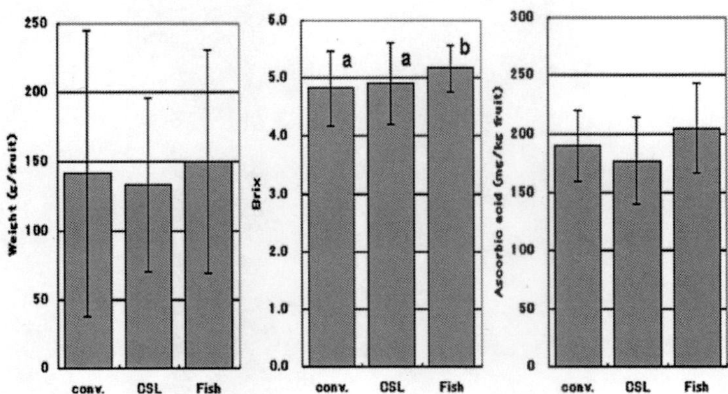

Figure 8. Tomato growth in the conventional and organic hydroponic systems. The growth and quality of tomato plants grown using the organic hydroponics approach with corn steep liquor (CSL) or fish-based soluble (Fish) fertilizer as the nitrogen source, and in a conventional hydroponics system with inorganic fertilizer (conv.) were measured. Values represent means and standard deviations of fruit weight (n¼24) and of the Brix and ascorbic acid data (n¼32 fruits).

Biofilms of microorganisms also developed on the root hairs of roots submerged in the solutions containing CSL or fish-based fertilizer, but neither biofilms nor root hairs were observed in the conventional hydroponic system (Fig. 9).

The concentrations of nitrate ion in the hydroponic solutions from both organic systems were decreased gradually and the nitrate ion could not be detected in the solution 14 days after the start of cultivation (Fig. 10). In contrast, nitrate ion had been detected in the solution of the conventional system since then.

DISCUSSION

The addition of only fish-based soluble fertilizer without inoculum in distilled water resulted in the generation of ammonium but no nitrate (Fig. 1).

Figure 9. Tomato growth in the conventional and organic hydroponic systems. The growth of tomato plants grown with three kinds of fertilizer was compared: (a) conventional inorganic fertilizer, and (b, c) microbial inoculum combined with (b) corn steep liquor and (c) fish-based soluble fertilizer. Pho-

tographs show (i) the tomato plants just before harvesting, (ii) the tomato roots on June 15, 2006 during cultivation, and (iii) the roots submerged in the hydroponic solution on June 27, 2006. Tomato roots were observed submerged in the hydroponic solution under a stereomicroscope. (a(iv)) Roots of tomato grown in the conventional hydroponic system with inorganic fertilizer had no root hairs. (c(iv)) Large numbers of root hairs developed on the roots of tomato grown in the organic hydroponic system with fish-based soluble fertilizer. The hairs were also covered with a biofilm.

In contrast, using inoculum such as nursery or field soil or bark compost resulted in nitrate generation. These results suggest that saprophytic microorganisms contained in fish-based soluble fertilizer cannot generate nitrate from its organic nitrogen, and that it is necessary to add inoculum to generate nitrate. Interestingly, the addition of only organic fertilizer without inoculum in sea water also resulted in generation of nitrate. This indicates that sea water contains microorganisms capable of degrading organic fertilizer to produce nitrate.

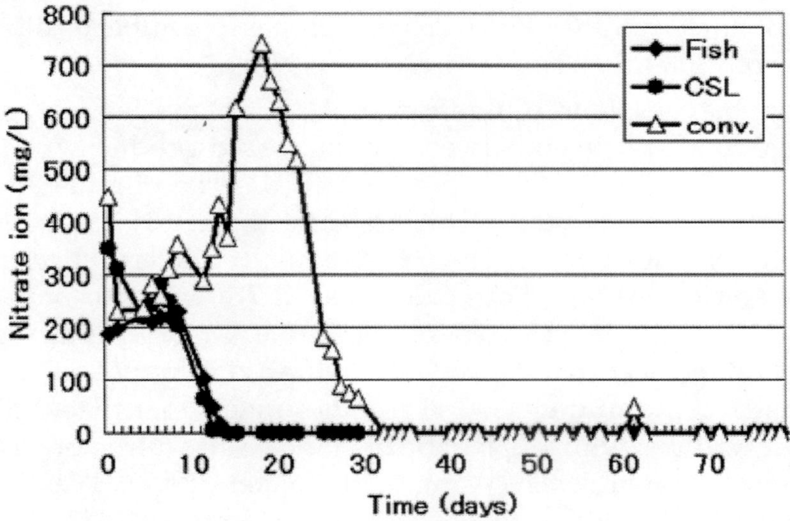

Figure 10. Nitrate concentration in the conventional and organic hydroponic solutions. The concentration of nitrate in the hydroponic solution with corn steep liquor fertilizer (CSL), fish-based soluble fertilizer (Fish), and conventional hydroponics with inorganic fertilizer (conv.) during tomato cultivation was measured.

Control of the amounts of both inoculum and organic fertilizer is important in order to generate nitrate in water. Too small an amount of inoculum (0.5 g/L bark compost) resulted in the generation of only ammonium when 0.25 g of CSL was added per 100mL (Fig. 2). Too much fish-based soluble fertilizer (2.5 g/L daily for seven days) also resulted in the generation of only ammonium (Fig. 3).

The growth of nitrifying bacteria, such as the obligate chemolithoautotrophs Nitrosomonas spp. And Nitrospira spp., is particularly inhibited by the presence of organic compounds (Jensen 1950; Quastel and Scholefield 1951; Rittenberg 1969; Smith and Hoare 1977; Krummel and Harms 1982; Takahashi et al. 1992; Stutte 1996; Xu et al. 2000; Tomiyama et al. 2001). Therefore, to culture microorganisms capable of generating nitrate in water, it is necessary to add sufficient inoculum that contains nitrifying microorganisms and to restrict the amount of organic fertilizer that is added to the water.

However, it was not clear why adding 0.5 g/L bark compost generated no nitrate when 0.25 g of CSL was added per 100 mL, but generated nitrate when 0.25 g of fish-based soluble fertilizer was added per 100 mL.

This may relate to differences in the form and biodegradability of the organic nitrogen between animal-based products (fish-based soluble fertilizer) and plant products (CSL). The nitrate-generation efficiencies of the organic fertilizers were not clearly related to their C/N ratios. However, high-C/N organic fertilizers, such as low-grade spirits distilled from sake lees (11.78) and rice bran (18.1), were not appropriate for generating nitrate. Organic fertilizers with a C/N ratio of less than 11, such as soybean curd refuse (10.43), were mineralized into nitrate. These results suggest that the addition of organic fertilizer with a C/N ratio of 11 may cause nitrogen starvation of the microorganisms (Wang and Bakken 1997). We therefore suggest that organic fertilizers with a C/N ratio of <11 should be used to generate nitrate in an organic hydroponic system, or that a supplemental nitrogen source must be provided to decrease the C/N ratio. The microbial culture solution was suitable for organic hydroponics (Fig. 4a). In tomato experiment, the nitratenitrogen concentration was increased by the addition of fish-based soluble fertilizer. In contrast, the addition of fish-based soluble fertilizer in the conventional inorganic hydroponic solution with no inoculum caused the death of all the tomato seedlings and generated only ammonium (Fig. 4b). As we mentioned in the Introduction, excessive ammonium can be deleterious to the growth of vegetables, such as tomato, and many vegetable crops prefer nitrate (Ikeda and Osawa 1981; Strayer et al. 1997; Miyata and Ikeda 2005; Cruz et al. 2006).

Interestingly, lettuce seedlings grew well in the conventional inorganic hydroponic solution added with fishbased soluble fertilizer (Fig. 5c) as observed in the microbial culture solution (Fig. 5a). These results suggest that lettuce is more insusceptible to ammonium than tomato because lettuce is known as ammonium-philic plant (Tadano and Tanaka 1976; Ikeda and Osawa 1981). However, lettuce seedlings seriously wilted in the solution added with only fish-based soluble fertilizer (Fig. 5b) and only ammonium was detected in the solution. In previous studies for organic hydroponics, application of hydroponic solution, derived from organic fertilizer, with ammonium-nitrogen without nitrate-nitrogen significantly suppressed lettuce growth (Schwartzkopf and Stroup 1993; Atkin and Nichols 2004). These results suggest that supplying nitrate by microbial nitrification or addition of conventional inorganic fertilizer is required for the growth of lettuce when ammonium nutrient solution derived from organic fertilizer is used in hydroponic cultivation. For successful organic hydroponics, it will therefore be important to use a microbial culture solution, such as the ones in the present study that can generate nitrate from organic fertilizer, as the hydroponic solution.

The addition of oyster shell lime in the hydroponic solution can effectively supplement minor nutrients (Fig. 6). However, some leaf chlorosis occurs without this supplement, although the symptom of yellow leaf tips can be ameliorated by the addition of supplemental iron. These results suggest that the addition of oyster shell lime may be especially effective to eliminate iron deficiency. Shells such as those from oysters contain high levels of calcium carbonate (CaCO3) as their principal component. The solubility of CaCO3 in water is low (about 0.15 mmol/L at 25C). This may explain why malabsorption of potassium or magnesium after the addition of oyster shell lime was not observed in the lettuce and tomato growth experiments (Hewitt and Smith 1975).

The fresh weight of lettuce was significantly greater in the organic system than in the conventional system. In previous studies of organic fertilizer in hydroponics, lettuce dry weights were similar to or less than those with conventional inorganic fertilizer but fresh weights were significantly lower (Anderson and Schmidt 2001; Atkin and Nichols 2004). It is not clear why higher root dry weight was produced in the organic system. Root development is promoted by the presence of certain microorganisms (Compant et al. 2005), so we should study the relationship between the action of the microorganisms present in the hydroponic solution and the resulting effect on root development

It is not clear why the lettuce leaf nitrate content was significantly lower in the organic system than in the conventional system. It is assumed that it is attributed to larger growth of lettuce in the organic system (Fig. 7). In the hydroponic solution of the organic system, we only detected nitrate (235 mg/L nitrate); in the conventional inorganic hydroponic solution, the same amount of nitrogen was present, and comprised 1.8% ammoniumnitrogen and 98.2% nitrate-nitrogen. Nitrate ions at high concentrations in foods can harm the health of humans and animals (Blom-Zandstra 2008). These results suggest that the yield and quality of butterhead lettuce in the microorganism culture system were at least as good as those in the conventional inorganic chemical system.

It is interesting that in the tomato growth experiment, neither nitrate nor ammonium could be detected two weeks after the start of cultivation in the organic system (Fig. 8). In contrast, nitrate was detected up to one month after the start of cultivation in the conventional inorganic hydroponic system, despite the addition of the same amount of nitrogen. We found that biofilm collected from the surface of the roots was able to generate nitrate within a few days when supplied with organic fertilizer in water (data not shown). This suggests that the biofilm on the root surfaces was able to degrade the organic fertilizer into ammonium and then nitrate, which was then absorbed by the roots without diffusing into the hydroponic solution.

This may be why inorganic nitrogen disappeared from the hydroponic solution two weeks after the start of cultivation in the organic system. The forms or composition of the absorbed nitrogen, such as ammonium or nitrate, were unclear because neither nitrate nor ammonium was detected in the hydroponic solution. In future research, we should study the mechanism of absorption of nitrogen in these systems. The development of root hairs and biofilms on the roots submerged in the solutions is characteristic of the organic system (Fig. 9). It is known that few root hairs develop on roots submerged in a conventional inorganic hydroponic solution (Nakano et al. 2002). Little is known about the mechanism of root hair development (Michael 2001). The development of the root hairs in the organic system is therefore an important topic for future study. From the biofilm present on the tomato roots in the organic system, we were able to isolate Arthrobacter spp., Ralstonia spp., Mycobacterium spp., Bordetella spp., and Rhodoferax spp. by means of dilution plating, but they were not detected in our 16 S rRNA denaturinggradient gel electrophoresis analysis (data not shown). This indicates that these isolates may be not major components of the biofilms. If we try to recreate the generation of nitrate from organic fertilizer using single isolates from biofilms, it will be necessary to isolate the major microorganisms responsible for the degradation of the organic fertilizer and to find an appropriate combination of those isolates. This will be an important issue for future study, since the development of a reliable organic hydroponic system will require consistency in the inoculum source.

Organic hydroponics appears to be a suitable medium for observations of the interactions between rhizobacteria and roots. This interaction occurs in soil but is difficult to observe without disruption of the rhizosphere structure. In conventional hydroponics, it is easy to observe roots, but there is little or no interaction between microorganisms and the roots because there are few microorganisms in the hydroponic solution. In contrast, our organic hydroponic method allows observation of these interactions, at any time, without disturbing the roots or microorganisms (Fig. 9). Biofilms are an important subject in studies of bacterial communication and interactions, and the possibility that the bacterial community in a biofilm interacts with plant roots is intriguing (Ramey et al. 2004).

The results of the tomato growth experiments suggest that the yield and quality of the tomato fruits produced by the organic fertilizer system were not inferior to those produced by the conventional system. In previous studies of organic hydroponics, it was difficult to come up with an organic alternative to the conventional inorganic solutions used for growing tomatoes (Atkin and Nichols 2004; Miyata and Ikeda 2005). Inconvenient procedures to treat the organic fertilizer have been used, such as conducting the ammonification and nitrification reactions in separate bioreactor tanks, followed by eliminating the organic compounds and microorganisms from the mineralized solution before it was supplied to the plants (MacElroy and Bredt 1984; Tako et al. 1995; Garland et al. 1997; Atkin and Nichols 2004; Jewell and Kubota 2005; Miyata and Ikeda 2005). Furthermore, it was impossible to add organic fertilizer to the hydroponic solution during cultivation. In contrast, our method shows that it is possible to supply sufficient nutrients by adding organic fertilizer directly to the hydroponic solution during cultivation. It is therefore possible to use organic fertilizer in hydroponics.

In conventional hydroponics, microbial contamination of hydroponic solutions often results in root disease (Stanghellini and Rasmussen 1994; Vestergard 1994); this is why hydroponics researchers have traditionally regarded all microorganisms as deleterious. However, it is difficult and expensive to maintain axenic cultivation conditions in conventional hydroponics. In contrast, our results demonstrate that our novel technique is a feasible non-axenic alternative to conventional hydroponics. By culturing suitable non-pathogenic soil organisms, our cultivation system might be simplified and made less expensive. As seen in the tomato results, the ability to add organic fertilizer directly to the hydroponic solution is very convenient.

CONCLUSION

Bark compost, field soil, nursery soil, and sea water are appropriate sources of microorganisms to generate nitrate from organic fertilizer in water. About 5 g/L of bark compost inoculum is needed to mineralize organic fertilizer into nitrate; the addition of too little inoculum will prevent sufficient nitrate production. Nitrate was generated only in containers that received 0.5 g/L of fishbased soluble fertilizer daily for seven days; application rates of 2.5 g/L daily or higher for seven days produced only ammonium. The culture solution of microorganisms could be used as the hydroponic solution in our organic hydroponics system. The yield and quality of butterhead lettuce and tomato fruits were not significantly different from those in the conventional hydroponics system. Our method, which adds organic fertilizer directly into the hydroponic solution during cultivation, is convenient and practical. Unlike in previous studies of organic hydroponics, the organic fertilizer does not require pre-processing before it can be incorporated in the system.

ACKNOWLEDGMENTS

The authors would like to thank H. Iwakiri, M. Kohno, S. Kataoka, E. Fujii, H. Kawashima, Y. Nakano, K. Honda, K. Nishi, F. Terami, M. Kubota, J. Ohnishi, Y. Iida, and many other staff members of Japan's National Agricultural Research Organization for their support.

REFERENCES

1. Anderson RG, Schmidt LS 2001: Nutrient analysis of commercial organic fertilizers for greenhouse vegetable production. Hortscience, 36, 503.
2. Atkin K, Nichols MA 2004: Organic hydroponics. Acta Hort., 648, 121–127.
3. Blom-Zandstra M 2008: Nitrate accumulation in vegetables and its relationship to quality. Ann. Appl. Biol., 115, 553–561.
4. Compant S, Duffy B, Nowak J, Cle´ment C, Ait Barka E 2005: Use of plant growth-promoting bacteria for biocontrol of plant diseases: principles, mechanisms of action, and future prospects. Appl. Environ. Microbiol., 71, 4951–4959.
5. Cruz C, Bio AF, Domı´nguez-Valdivia MD, Aparicio-Tejo PM, Lamsfus C, Martins-Louc¸a˜o MA 2006: How does glutamine synthetase activity determine plant tolerance to ammonium? Planta, 223, 1068–1080.
6. Ehret DJ, Menzies JG, Helmer T 2005: Production and quality of greenhouse roses in recirculating nutrient systems. Sci. Hortic., 106, 103–113.
7. Garland JL, Mackoqiak JL 1990: Utilization of the Water Soluble Fraction of Wheat Straw as a Plant Nutrient Source, NASA Technical Memorandum 107544, NASA,
8. Huntsville, AL. Garland JL, Mackoqiak JL, Sager JC 1993: Hydroponic Crop Production Using Recycled Nutrients from Inedible Crop Residues, SAE Technical Paper 932173. SAE, Warrendale, PA.
9. Garland JL, Mackowiak CL, Strayer RF, Finger BW 1997: Integration of waste processing and biomass production systems as part of the KSC Breadboard project. Adv. Space Res., 20, 1821–1826.
10. Givan CV 1979: Metabolic detoxification of ammonia in tissues of higher plants. Phytochemistry, 18, 375–382.

11. Hagin J, Lowengart A 1996: Fertigation for minimizing environmental pollution by fertilizers. Fert. Res., 43, 5–7.

12. Hewitt EJ, Smith TA 1975: Plant Mineral Nutrition, English Universities Press, London.

13. Ikeda H, Osawa T 1981: Nitrate- and ammonium-N absorption by vegetables from nutrient solution containing ammonium nitrate and the resultant change of solution pH.

14. J. Jpn. Soc. Hort. Sci., 50, 225–230. Jensen HL 1950: Effect of organic compounds on Nitrosomonas. Nature, 165, 974.

15. Jewell B, Kubota C 2005: Challenges of organic hydroponic production of strawberries (Fragariaananassa). HortScience, 40, 1010–1011.

16. Krummel A, Harms H 1982: Effect of organic matter on growth and cell yield of ammonia-oxidizing bacteria. Arch. Microbiol., 133, 50–54.

17. Lee JG, Lee BY, Lee HJ 2006: Accumulation of phytotoxic organic acids in reused nutrient solution during hydroponic cultivation of lettuce (Lactuca sativa L.). Sci. Hortic., 110, 119–128.

18. MacElroy RD, Bredt J 1984: Current concepts and future direction of CELSS. Adv. Space Res., 4, 221–229.

19. Mackowiak CL, Garland JL, Strayer RF, Finger BW, Wheeler RM 1996: Comparison of aerobically-treated and untreated crop residue as a source of recycled nutrients in a recirculating hydroponic system. Adv. Space Res., 18, 282–287.

20. Michael G 2001: The control of root hair formation: suggested mechanisms. J. Plant Nutr. Soil Sci., 164, 111–119.

21. Miyata H, Ikeda H 2005: Using liquid methane-fermentation product and livestock urine as liquid fertilisers for horticultural production. In Management of Nutrients and Solid Media in Drip Fertigation, Eds. K Watanabe, Y Uehara, T Kimura, H Ikeda, pp 119—155, Hakuyusha Co., Ltd, Tokyo (in Japanese).

22. Nakano Y, Watanabe S, Okano K, Tatsumi J 2002: The influence of growing temperatures on activity and structure of tomato roots hydroponically grown in wet atmosphere or in solution. J. Jpn. Soc. Hort. Sci., 71, 683–690.

23. Purich GS, Barker AV 1967: Structure and function of tomato leaf chloroplasts during ammonium toxicity. Plant Physiol., 42, 1229–1238.

24. Quastel JH, Scholefield PG 1951: Biochemistry of nitrification in soil. Bacteriol. Rev., 15, 1–53.

25. Ramey BE, Koutsoudis M, von Bodman SB, Fuqua C 2004: Biofilm formation in plant-microbe associations. Curr. Opin. Microbiol., 7, 602–609.

26. Rittenberg SC 1969: The roles of exogenous organic matter in the physiology of chemolithotrophic bacteria. Adv. Microb. Physiol., 3, 159–196. Schwartzkopf SH, Stroup TL 1993: Anaerobically-Processed Waste as a Nutrient Source for Higher Plants in a Controlled Ecological Life Support System, SAE Technical Paper 932248. SAE, Warrendale, PA.

27. Shinohara M 2006: Hydroponics with organic fertilizers—a method for building an ecological system of microorganisms in culture liquid by a parallel mineralization method. Agric. Hortic., 81, 753–764 (in Japanese).

28. Shinohara M, Uehara Y, Kouno M, Iwakiri H 2007: Method for producing biomineral-containing substance and organic hydroponics method. Japanese Patent Application Laid-Open No. 2007-119260.

29. Smith AJ, Hoare DS 1977: Specialist phototrophs, lithotrophs, and methylotrophs: a unity among a diversity of prokaryotes?. Bacteriol Rev., 41, 419–448.

30. Stanghellini ME, Rasmussen SL 1994: Identification and origin of plant pathogenic microorganisms in recirculating nutrient solutions. Adv. Space Res., 14, 349–355.

31. Strayer RF, Finger BW, Alazraki MP 1997: Evaluation of an anaerobic digestion system for processing CELSS crop residues for resource recovery. Adv. Space Res., 20, 2009–2015.

32. Stutte GW 1996: Nitrogen dynamics in the CELSS breadboard facility at Kennedy Space Center. Life Support Biosph. Sci., 3, 67–74.

33. Tadano T, Tanaka A 1976: Comparison of adaptability to ammonium and nitrate among crop plants. (Part 1) Selective absorption between and responses to ammonium and nitrate of crop plants during early stage. Studies on the comparative plant nutrition. J. Sci. Soil and Manure Japan., 47, 321–328 (In Japanese).

34. Takahashi R, Kondo N, Usui K, Kanehira T, Shinohara M, Tokuyama T 1992: Pure isolation of a new chemoautotrophic ammonia-oxidizing bacterium on gellan gum plate. J. Ferment. Bioeng., 74, 52–54.

35. Tako Y, Saito T, Tani A, Terai M, Nitta K 1995: A study on utilization of processed organic waste solution for plant cultivation in closed ecological systems. An experimental study on improvement of suitability of waste wet-oxidized solution for use as plant nutrient solution. CELSS J, 7, 19–26.

36. Tomiyama H, Ohshima M, Ishii S, Satoh K, Takahashi R, Isobe K, Iwano H, Tokuyama T 2001: Characteristics of newly isolated nitrifying bacteria from rhizoplane of paddy rice. Microb. Environ., 16, 101–108.

37. Trebst AV, Losada M, Arnon DI 1960: Photosynthesis by isolated chloroplasts. J. Biol. Chem., 235, 840–844.

38. Vestergard B 1994: Establishing and maintaining specific pathogen free (SPF) conditions in aqueous solutions using ozone. Adv. Space Res., 14, 387–393.

39. Wang JG, Bakken LR 1997: Competition for nitrogen during mineralization of plant residues in soil: microbial response to C and N availability. Soil Biol. Biochem., 29, 163–170.

40. Wheeler RM, Mackowiak CL, Sager JC, Knott WM, Hinkle CR 1990: Potato growth and yield using nutrient film technique (NFT). Am. Potato J., 67, 177–187.

41. Xu Z, Zheng S, Yang G, Zhang Q, Wang L 2000: Nitrification inhibition by naphthalene derivatives and its relationship with copper. Bull. Environ. Contam. Toxicol., 64, 542–549.

Citations

CHAPTER 1

Hala Kandil, Nadia Gad Effects Of Inorganic And Organic Fertilizers On Growth And Production Of Brocoli (Brassica Oleracea L.) factori. soilscience.ro/index.php/fspdzt/article/view/160/91.

CHAPTER 2

Bimova PAVLA, Robert POKLUDA, Influence of Alternative Organic Fertilizers on the Antioxidant Capacity in Head Cabbage and Cucumber, Print ISSN 0255-965X; Electronic ISSN 1842-4309

CHAPTER 3

Zhao Chen 1 and Xiuping Jiang, Microbiological Safety of Chicken Litter or Chicken Litter-Based Organic Fertilizers: A Review, doi:10.3390/agriculture4010001.

CHAPTER 4

Eduardo Fávero Caires; Helio Antonio Wood Joris; Susana Churka; Renato Zardo Filho, Performance of maize landrace under no-till as affected by the organic and mineral fertilizers, http://dx.doi.org/10.1590/S1516-89132012000200006

CHAPTER 5

W. R. Alharbi, Natural Radioactivity and Dose Assessment for Brands of Chemical and Organic Fertilizers Used in Saudi Arabia , DOI: 10.4236/jmp.2013.43047.

CHAPTER 6

Hou, Y.X., Hu, X.J., Yan, W.T., Zhang, S.H.and Niu, L.B.(2013) Effect of organic fertilizers used in sandy soil on the growth of tomatoes. Agricultural Sciences, 4, 31-34. doi: 10.4236/as.2013.45B006.

CHAPTER 7

Edwin Mwiti Mutegi, James Biu Kung'u, Mucheru-Muna, Pypers Pieter, Daniel Njiru Mugendi, Complementary Effects of Organic and Mineral Fertilizers on Maize Production in the Smallholder Farms of Meru South District, Kenya, DOI: 10.4236/as.2012.32026.

CHAPTER 8

Alessandra Trinchera, Carlos Mario Rivera, Andrea Marcucci and Elvira Rea (2013). Biomass Digestion to Produce Organic Fertilizers: A Case-Study on Digested Livestock Manure, Biomass Now - Cultivation and Utilization, Dr. Miodrag Darko Matovic (Ed.), ISBN: 978-953-51-1106-1, InTech, DOI: 10.5772/53869.

CHAPTER 9

Mohd Hafiz Ibrahim, Hawa Z. E. Jaafar, Ehsan Karimi and Ali Ghasemzadeh , Impact of Organic and Inorganic Fertilizers Application on the Phytochemical and Antioxidant Activity of Kacip Fatimah (Labisia pumila Benth), doi:10.3390/molecules180910973.

CHAPTER 10

M.K.C. Sridhar, G.O. Adeoye, and O.O. AdeOluwa, "Alternate Nitrogen Amendments for Organic Fertilizers," TheScientificWorldJOURNAL, vol. 1, pp. 142-147, 2001. doi:10.1100/tsw.2001.454

CHAPTER 11

Malancha Roy1, Sukalpa Karmakar1, Anupam Debsarcar2, Pradip K Sen3 and Joydeep Mukherjee1*Application of rural slaughterhouse waste as an organic fertilizer for pot cultivation of solanaceous vegetables in India .ijrowa.com/content/2/1/6

CHAPTER 12

Nagavallemma KP, Wani SP, Stephane Lacroix, Padmaja VV, Vineela C, Babu Rao M and Sahrawat KL Vermicomposting: Recycling Wastes into Valuable Organic Fertilizer ejournal.icrisat.org/agroecosystem/v2i1/v2i1vermi.pdf

CHAPTER 13

Artur Granstedt[1] & Lars Kjellenberg[2], Long-Term Field Experiment in Sweden: Effects of Organic and Inorganic Fertilizers on Soil Fertility and Crop Quality

CHAPTER 14

Makoto Shinohara , Chihiro Aoyama , Kazuki Fujiwara , Atsunori Watanabe , Hiromi Ohmori , Yoichi Uehara & Masao Takano (2011) Microbial mineralization of organic nitrogen into nitrate to allow the use of organic fertilizer in hydroponics, Soil Science and Plant Nutrition, 57:2, 190-203, DOI: 10.1080/00380768.2011.554223

INDEX

V

W